SCHAUM'S *Easy* OUTLINES

CALCULUS

Other Books in Schaum's Easy Outline Series include:

Schaum's Easy Outline: College Algebra
Schaum's Easy Outline: College Physics
Schaum's Easy Outline: Statistics
Schaum's Easy Outline: Programming in C++
Schaum's Easy Outline: College Chemistry
Schaum's Easy Outline: French
Schaum's Easy Outline: Spanish
Schaum's Easy Outline: German
Schaum's Easy Outline: Organic Chemistry

SCHAUM'S *Easy* OUTLINES

CALCULUS

BASED ON SCHAUM'S
Outline of Differential and Integral Calculus
BY FRANK AYRES, JR. AND
ELLIOT MENDELSON

ABRIDGEMENT EDITOR:
GEORGE J. HADEMENOS

SCHAUM'S OUTLINE SERIES
McGRAW-HILL

New York   San Francisco   Washington, D.C.   Auckland   Bogotá
Caracas   Lisbon   London   Madrid   Mexico City   Milan   Montreal
New Delhi   San Juan   Singapore   Sydney   Tokyo   Toronto

FRANK AYRES, Jr., was formerly Professor and Head of the Department of Mathematics at Dickinson College, Carlisle, Pennsylvania.

ELLIOTT MENDELSON is Professor of Mathematics at Queens College.

GEORGE J. HADEMENOS has taught at the University of Dallas and performed research at the University of California at Los Angeles and the University of Massachusetts Medical Center. He earned a B.S. degree in Physics from Angelo State University and the M.S. and Ph.D. degrees in Physics from the University of Texas at Dallas.

3 4 5 6 7 8 9 10 11 12 13 14 15 DOC DOC   0 9 8 7 6 5 4 3 2 1 0

ISBN 0-07-052710-5

Sponsoring Editor: Barbara Gilson
Production Supervisor: Tina Cameron
Editing Supervisor: Maureen B. Walker

McGraw-Hill

A Division of The McGraw-Hill Companies

Contents

Chapter 1
FUNCTIONS, SEQUENCES, LIMITS, AND CONTINUITY

IN THIS CHAPTER:

✔ *Function of a Variable*
✔ *Graph of a Function*
✔ *Infinite Sequence*
✔ *Limit of a Sequence*
✔ *Limit of a Function*
✔ *Right and Left Limits*
✔ *Theorems on Limits*
✔ *Continuity*
✔ *Solved Problems*

Function of a Variable

A **function** is a rule that associates, with each value of a variable x in a certain set, exactly one value of another variable y. The variable y is then called the **dependent variable**, and x is called the **independent variable**. The set from which the values of x can be chosen is called the **domain** of the function. The set of all the corresponding values of y is called the **range** of the function.

Example 1.1 The equation $x^2 - y = 10$, with x the independent variable, associates one value of y with each value of x. The function can be calculated with the formula $y = x^2 - 10$. The domain is the set of all real numbers. The same equation, $x^2 - y = 10$, with y taken as the independent variable, sometimes associates two values of x with each value of y. Thus, we must distinguish two functions of y:

$$x = \sqrt{10+y} \text{ and } x = -\sqrt{10+y}$$

The domain of both of these functions is the set of all y such that $y \geq -10$, since $\sqrt{10+y}$ is not a real number when $10 + y < 0$.

If a function is denoted by a symbol f, then the expression f(b) denotes the value obtained when f is applied to a number b in the domain of f. Often a function is defined by giving the formula for an arbitrary value f(x). For example, the formula $f(x) = x^2 - 10$ determines the first function mentioned in Example 1.1. The same function also can be defined by an equation like $y = x^2 - 10$.

Examples 1.2

(1) If $f(x) = x^3 - 4x + 2$, then
$$f(1) = (1)^3 - 4(1) + 2 = 1 - 4 + 2 = -1$$
$$f(-2) = (-2)^3 - 4(-2) + 2 = -8 + 8 + 2 = 2$$
$$f(a) = a^3 - 4a + 2$$

(2) The function $f(x) = 18x - 3x^2$ is defined for every number x; that is, without exception, $18x - 3x^2$ is a real number whenever x is a real number. Thus, the domain of the function is the set of all real numbers.

(3) The area A of a certain rectangle, one of whose sides has length x, is given by $A = 18x - 3x^2$. Here, both x and A must be positive. By completing the square, we obtain $A = -3(x-3)^2 + 27$. In order to have $A > 0$, we must have $3(x-3)^2 < 27$, which limits x to values below 6; hence, $0 < x < 6$. Thus, the function determining A has the open interval (0, 6) as its domain. From Figure 1-1, we see that the range of the function is the interval (0, 27].

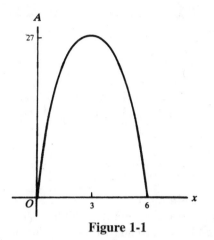

Figure 1-1

Graph of a Function

The graph of a function f is the graph of the set of points on the plane (x, y) satisfying the equation y = f(x).

Examples 1.3

(1) Consider the function f(x) = |x|. Its graph is the graph of the equation y = |x|, shown in Figure 1-2.

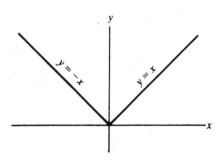

Figure 1-2

Notice that f(x) = x when x ≥ 0, whereas f(x) = -x when x ≤ 0. The domain of f consists of all real numbers, but the range is the set of all nonnegative real numbers.

(2) The formula g(x) = 2x + 3 defines a function g. The graph of this function is the graph of the equation y = 2x + 3, which is the straight line with slope 2 and y intercept 3. The set of all real numbers is both the domain and range of g.

Infinite Sequence

An **infinite sequence** is a function whose domain is the set of positive integers. For example, when n is given in turn the values 1, 2, 3, 4, ..., the function defined by the formula $1/(n + 1)$ yields the sequence 1/2, 1/3, 1/4, 1/5, ... The sequence is called an infinite sequence to indicate that there is no last term.

By the **general** or **nth term** of an infinite sequence, we mean a formula s_n for the value of the function determining the sequence. The infinite sequence itself is often denoted by enclosing the general term in braces, as in $\{s_n\}$, or by displaying the first few terms of the sequence. For example, the general term s_n of the sequence in the preceding paragraph is $1/(n + 1)$, and that sequence can be denoted by $\{1/(n + 1)\}$ or by 1/2, 1/3, 1/4, 1/5, ...

Limit of a Sequence

If the terms of a sequence $\{s_n\}$ approach a fixed number c as n gets larger and larger, we say that c is the **limit** of the sequence, and we write either $a_n \to c$ or $\lim_{n \to +\infty} a_n = c$. This means that $|a_n - c| < \varepsilon$, no matter how small $\varepsilon > 0$ is chosen to be.

Example 1.4 Consider the sequence

$$1, 3/2, 5/3, 7/4, 9/5, \ldots, 2-1/n, \ldots \tag{1.1}$$

whose terms are plotted on the coordinate system in Figure 1-3.

Figure 1-3

As *n* increases, consecutive points cluster toward the point 2 in such a way that the distance of the points from 2 eventually becomes less than any positive number that might have been preassigned as a measure of closeness, however small. (For example, the point 2 - 1/1001 = 2001/1001 and all subsequent points are at a distance less than 1/1000 from 2 [that is, ε = 1/1000], the point 20 000 001/10 000 001 and all subsequent points are at a distance less than 1/10 000 000 from 2 [that is, ε = 1/10 000 000], and so on.) Hence,

$$\{2 - 1/n\} \rightarrow 2 \qquad \text{or} \qquad \lim_{n \rightarrow +\infty} (2 - 1/n) = 2$$

The sequence (1-1) does not contain its limit 2 as a term. On the other hand, the sequence 1, 1/2, 1, 3/4, 1, 5/6, 1, . . . has 1 as limit, and every odd-numbered term is 1. Thus, a sequence having a limit may or may not contain that limit as a term.

Many sequences do not have a limit. For example, the sequence $\{(-1)^n\}$, that is -1, 1, -1, 1, -1, 1, . . . keeps alternating between -1 and 1 and does not get closer and closer to any fixed number.

Limit of a Function

If f is a function, then we say that $\lim_{x \rightarrow a} f(x) = A$ where A < ∞, if the value of f(x) gets arbitrarily close to A as x gets closer and closer to a, that is, the distance between them is small.

Example 1.5 The $\lim_{x \rightarrow 3} x^2 = 9$, since x^2 gets arbitrarily close to 9 as x approaches as close as one wishes to 3.

The definition can be stated more precisely as follows:

DEFINITION: $\lim\limits_{x \to a} f(x) = A$ if and only if, for any chosen positive number, ε, however small, there exists a positive number δ such that, whenever $0 < |x - a| < \delta$, then $|f(x) - A| < \varepsilon$. This can be illustrated in Figure 1-4.

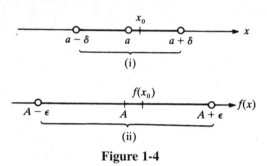

Figure 1-4

After ε has been chosen [that is, after interval (ii) has been chosen], then δ can be found [that is, interval (i) can be determined] so that, wherever $x \neq a$ is on interval (i) say at x_0 then $f(x)$ is on interval (ii), at $f(x_0)$.

Notice the important fact that whether or not $\lim\limits_{x \to a} f(x) = A$

is true does not depend upon the value of $f(x)$ when $x = a$. In fact, $f(x)$ need not even be defined when $x = a$.

Example 1.6

$$\lim_{x \to 2} \frac{x^2 - 4}{x - 2} = 4$$

although $(x^2 - 4)/(x - 2)$ is not defined when $x = 2$. Since

$$\frac{x^2 - 4}{x - 2} = \frac{(x - 2)(x + 2)}{(x - 2)} = x + 2$$

then we see that $(x^2 - 4)/(x - 2)$ approaches 4 as x approaches 2.

Example 1.7 Let us use the precise definition to show that

$$\lim_{x \to 2} (x^2 + 3x) = 10$$

Let $\varepsilon > 0$ be chosen. We must produce a $\delta > 0$ such that, whenever $0 < |x - 2| < \delta$ then $|(x^2 + 3x) - 10| < \varepsilon$. First we note that

$$|(x^2 + 3x) - 10| = |(x - 2)^2 + 7(x - 2)| \leq |x - 2|^2 + 7|x - 2|$$

where $|x - 2| < \delta$. Also, if $0 < \delta \leq 1$, then $\delta^2 \leq \delta$. Hence, if we take δ to be the minimum of 1 and $\varepsilon / 8$, then, whenever $0 < |x - 2| < \delta$,

$$|(x^2 + 3x) - 10| < \delta^2 + 7\,\delta \leq \delta + 7\,\delta = 8\delta \leq \varepsilon$$

Right and Left Limits

By $\lim\limits_{x \to a-} f(x) = A$ where $A < \infty$, we mean that $f(x)$ approaches A as x approaches a through values less than a, that is, as x approaches a *from the left*. Similarly, $\lim\limits_{x \to a+} f(x) = A$ means that $f(x)$ approaches A as x approaches a through values greater than a, that is, as x approaches a *from the right*. The statement $\lim\limits_{x \to a} f(x) = A$ is equivalent to the conjunction of the two statements $\lim\limits_{x \to a-} f(x) = A$ and $\lim\limits_{x \to a+} f(x) = A$.

For A to be the limit of the function $f(x)$ as x a, it must be unique and finite. The existence of the limit from the left does not imply the existence of the limit from the right, and conversely. When a function f is defined on only one side of a point a, then $\lim\limits_{x \to a} f(x)$ refers to the one--sided limit, if it exists.

Example 1.8 The function $f(x) = \sqrt{x}$; then f is defined only to the right of zero. Hence, $\lim\limits_{x \to 0} \sqrt{x} = \lim\limits_{x \to 0+} \sqrt{x} = 0$. Of course, $\lim\limits_{x \to 0+} \sqrt{x}$ does not exist, since $\sqrt{x}$ is not defined when $x < 0$.

Theorems on Limits

The following theorems on limits are listed for future reference:

Theorem 1-1:　If $f(x) = c$, a constant, then $\lim_{x \to a} f(x) = c$.

If $\lim_{x \to a} f(x) = A$ and $\lim_{x \to a} g(x) = B$ where $A, B < \infty$, then:

Theorem 1-2:　$\lim_{x \to a} k\, f(x) = kA$, where k is any constant

Theorem 1-3:　$\lim_{x \to a} [f(x) \pm g(x)] = \lim_{x \to a} f(x) \pm \lim_{x \to a} g(x) = A \pm B$

Theorem 1-4:　$\lim_{x \to a} [f(x) \cdot g(x)] = \lim_{x \to a} f(x) \cdot \lim_{x \to a} g(x) = A \cdot B$

Theorem 1-5:　$\lim_{x \to a} \dfrac{f(x)}{g(x)} = \dfrac{\lim f(x)}{\lim g(x)} = \dfrac{A}{B}$, provided $B \neq 0$

Theorem 1-6:

$\lim_{x \to a} \sqrt[n]{f(x)} = \sqrt[n]{\lim_{x \to a} f(x)} = \sqrt[n]{A}$, provided $\sqrt[n]{A}$ is a real number

Continuity

A function $f(x)$ is called **continuous** if it is continuous at every point of its domain. A function $f(x)$ is continuous at $x = x_0$ if $f(x_0)$ is defined; $\lim_{x \to x_0} f(x)$ exists; and, $\lim_{x \to x_0} f(x) = f(x_0)$.

A function f is said to be continuous on a closed interval $[a, b]$ if the function that restricts f to $[a, b]$ is continuous at each point of $[a, b]$; in other words, we ignore what happens to the left of a and to the right of b. A function f(x) is **discontinuous** at $x = x_0$ if one or more of the conditions for continuity fails there.

Example 1.9. Determine the continuity of:

(a) $f(x) = \dfrac{1}{x-2}$;   (b) $f(x) = \dfrac{x^2-4}{x-2}$

(a) This function is discontinuous at $x = 2$ because $f(2)$ is not defined (has zero as denominator) and because $\lim\limits_{x \to 2} f(x)$ does not exist (equals ∞).

The function is, however, continuous everywhere except at $x = 2$, where it is said to have an **infinite discontinuity.** See Figure 1-5.

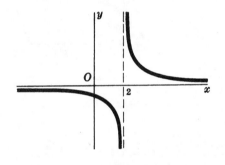

Figure 1-5

(b) This function is discontinuous at $x = 2$ because $f(2)$ is not defined (both numerator and denominator are zero), however $\lim\limits_{x \to 2} f(x) = 4$.

The discontinuity here is called **removable** since it may be removed by redefining the function $f(x)$, that is, reducing it algebraically so as to obtain a function $g(x)$ which is continuous at $x = 2$:

$$g(x) = \frac{x^2-4}{x-2} = \frac{(x+2)(x-2)}{x-2} = x+2$$

$$g(2) = 2 + 2 = 4$$

The discontinuity in part (a) cannot be so removed because the limit also does not exist.

The graphs of

$$f(x)=\frac{x^2-4}{x-2}\,;\ g(x)=x+2$$

are identical except at $x = 2$, where the former has a "hole" (see Figure 1-6). Removing the discontinuity consists simply of filling the "hole."

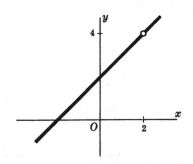

Figure 1-6

Properties of Continuous Functions

The theorems on limits lead readily to theorems on continuous functions. In particular, if $f(x)$ and $g(x)$ are continuous at $x = a$, so also are $f(x)\pm g(x)$, $f(x){\cdot}g(x)$, and $f(x)/g(x)$, provided in the latter that $g(a) \neq 0$. Hence, polynomials in x are continuous everywhere whereas rational functions of x are continuous everywhere except at values of x for which the denominator is zero.

The property of continuous functions used here is:

Property 1.1: If $f(x)$ is continuous on the interval $a \leq x \leq b$ and if $f(a) \neq f(b)$, then for any number c between $f(a)$ and $f(b)$ there is at least one value of x, say $x = x_0$, for which $f(x_0) = c$ and $a \leq x_0 \leq b$.

Property 1.1 is also known as the **intermediate value theorem.** Figure 1-7 illustrates the two applications of this property, and Figure 1-8 shows that continuity throughout the interval is essential.

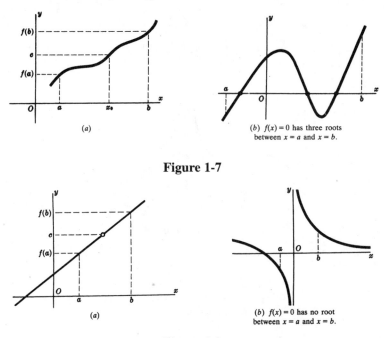

(a)

(b) $f(x) = 0$ has three roots between $x = a$ and $x = b$.

Figure 1-7

(a)

(b) $f(x) = 0$ has no root between $x = a$ and $x = b$.

Figure 1-8

Property 1.2: If f(x) is continuous on the interval $a \le x \le b$, then f(x) takes on a least value m and a greatest value M on the interval.

Consider Figure 1-8 through 1-10. In Figure 1-8, the function is continuous on $a \le x \le b$; the least value m and the greatest value M occur at x = c and x = d respectively, both points being within the interval. In Figure 1-9, the function is continuous on $a \le x \le b$; the least value occurs at the endpoint x = a, while the greatest value occurs at x = c within the interval. In Figure 1-10, there is a discontinuity at x = c, where a < c < b; the function has a least value at x = a but no greatest value.

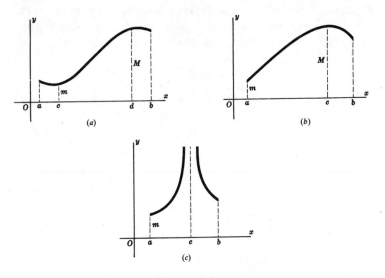

Figure 1-9

Property 1.3: If $f(x)$ is continuous on the interval $a \le x \le b$, and if c is any number between a and b and $f(c) > 0$, then there exists a number $\lambda > 0$ such that whenever $c - \lambda < x < c + \lambda$, then $f(x) > 0$.

This property is illustrated in Figure 1-10.

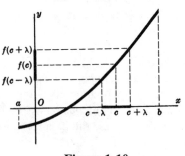

Figure 1-10

Solved Problems

Solved Problem 1.1 A rectangular plot requires 2000 ft of fencing to enclose it. If one of its dimensions is x (in feet), express its area y (in square feet) as a function of x, and determine the domain of the function.

Solution. Since one dimension is x, the other is

$\frac{1}{2}(2000-2x)=1000-x$ The area is then $y=x(1000-x$, and the do-

main of this function is $0 < x < 1000$.

Solved Problem 1.2 From each corner of a square of tin, 12 in on a side, small squares of side x (in inches) are removed, and the edges are turned up to form an open box (Figure SP1-1). Express the volume V of the box (in cubic inches) as a function of x, and determine the domain of the function.

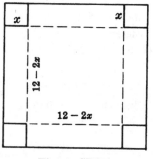

Figure SP1-1

Solution. The box has a square base of side 12 - 2x and a height of x. The volume of the box is then $V = x(12 - 2x)^2 = 4x(6 - x)^2$. The domain is the interval $0 < x < 6$. As x increases over its domain, V increases for a time and then decreases thereafter. Thus, among such boxes that may be constructed, there is one of greatest volume, say M. To determine M, it is necessary to locate the precise value of x at which V ceases to increase.

Solved Problem 1.3 If $f(x)=x^2+2x$, find $\dfrac{f(a+h)-f(a)}{h}$ and interpret the result.

Solution.

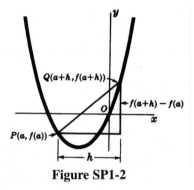

Figure SP1-2

On the graph of the function (Figure SP1-2), locate points P and Q whose respective abscissas are a and a + h. The ordinate of P is f(a), and that of Q is f(a + h). Then

$$\frac{f(a+h)-f(a)}{h} = \frac{\text{difference of ordinates}}{\text{difference of abscissas}} = \text{slope of PQ}$$

$$\frac{f(a+h)-f(a)}{h} = \frac{\left[(a+h)^2+2(a+h)\right]-(a^2+2a)}{h} = 2a+2+h$$

Solved Problem 1.4 Write the general term of each of the following sequences:

(a) $1, \dfrac{1}{3}, \dfrac{1}{5}, \dfrac{1}{7}, \dfrac{1}{9}, ...$

(b) $1, -\dfrac{1}{2}, \dfrac{1}{3}, -\dfrac{1}{4}, \dfrac{1}{5}, ...$

(c) $\dfrac{1}{2}, -\dfrac{4}{9}, \dfrac{9}{28}, -\dfrac{16}{65}, ...$

Solution.

(a)   The terms are the reciprocals of the odd positive integers. The general term is $\dfrac{1}{2n-1}$.

(b)   Apart from sign, these are the reciprocals of the positve integers. The general term is $(-1)^{n+1}\dfrac{1}{n}$ or $(-1)^{n-1}\dfrac{1}{n}$.

(c)   Apart from sign, the numerators are the squares of positive integers and the denominators are the cubes of these integers increased by 1. The general term is $(-1)^{n+1}\dfrac{n^2}{n^3+1}$.

Chapter 2
DIFFERENTIATION

IN THIS CHAPTER:

✔ *The Derivative*
✔ *Differentiation*
✔ *Differentiation Rules*
✔ *Inverse Functions*
✔ *The Chain Rule*
✔ *Higher Derivatives*
✔ *Implicit Differentiation*
✔ *Derivatives of Higher Order*
✔ *Solved Problems*

The Derivative

The change in the variable x between two values $x = x_0$ and $x = x_1$ in its domain is the **increment** Δx. Specifically, if $\Delta x = x_1 - x_0$, we may write $x_1 = x_0 + \Delta x$. If x changes by an increment Δx from an initial value $x = x_0$, then we write $x = x_0 + \Delta x$. In the same fashion, the change in a function $y = f(x)$ evaluated between $x = x_0$ and $x = x_0 + \Delta x$ is called the increment $\Delta y = f(x_0 + \Delta x) - f(x_0)$. Then, the quotient,

$$\frac{\Delta y}{\Delta x} = \frac{\text{change in y}}{\text{change in x}}$$

is called the **average rate of change** of the function on the interval between $x = x_0$ and $x = x_0 + \Delta x$.

Example 2.1 When x is given the increment $\Delta x = 0.5$ from $x_0 = 1$, the function $y = f(x) = x^2 + 2x$ is given the increment $\Delta y = f(1 + 0.5) - f(1)$ $= 5.25 - 3 = 2.25$. Thus, the average rate of change of y on the interval between $x = 1$ and $x = 1.5$ is

$$\frac{\Delta y}{\Delta x} = \frac{2.25}{0.5} = 4.5$$

The **derivative** of a function $y = f(x)$ with respect to x at the point $x = x_0$ is defined as

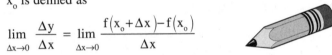

$$\lim_{\Delta x \to 0} \frac{\Delta y}{\Delta x} = \lim_{\Delta x \to 0} \frac{f(x_0 + \Delta x) - f(x_0)}{\Delta x}$$

provided the limit exists. This limit is also called the **instantaneous rate of change** of y with respect to x at $x = x_0$.

Example 2.2 Find the derivative of $y = f(x) = x^2 + 3x$ with respect to x at $x = x_0$. Use this to find the value of the derivative at (a) $x_0 = 2$ and (b) $x_0 = -4$.

$$f(x_0) = x_0^2 + 3x_0$$

$$f(x_0 + \Delta x) = (x_0 + \Delta x)^2 + 3(x_0 + \Delta x)$$
$$= x_0^2 + 2x_0 \Delta x + (\Delta x)^2 + 3x_0 + 3\Delta x$$
$$\Delta y = f(x_0 + \Delta x) - f(x_0) = 2x_0 \Delta x + 3\Delta x + (\Delta x)^2$$

$$\frac{\Delta y}{\Delta x} = \frac{f(x_0 + \Delta x) - f(x_0)}{\Delta x} = 2x_0 + 3 + \Delta x$$

The derivative at $x = x_0$ is

$$\lim_{\Delta x \to 0} \frac{\Delta y}{\Delta x} = \lim_{\Delta x \to 0} (2x_0 + 3 + \Delta x) = 2x_0 + 3$$

(a) At $x_0 = 2$, the value of the derivative is $2(2) + 3 = 7$.

(b) At $x_0 = -4$, the value of the derivative is $2(-4) + 3 = -5$.

In finding derivatives, it is customary to drop the subscript 0 and obtain the derivative of $y = f(x)$ **with respect to x** as

$$\lim_{\Delta x \to 0} \frac{\Delta y}{\Delta x} = \lim_{\Delta x \to 0} \frac{f(x + \Delta x) - f(x)}{\Delta x}$$

The derivative of $y = f(x)$ with respect to x may be indicated by any one of the symbols

$$\frac{d}{dx} y \qquad \frac{dy}{dx} \qquad D_x y' \qquad f'(x) \qquad \frac{d}{dx} f(x)$$

Differentiation

A function is said to be **differentiable** at a point $x = x_0$ if the derivative of the function exists at that point. Also, a function is said to be differentiable on an interval if it is differentiable at every point of the interval. The functions of elementary calculus are differentiable, except possibly at isolated points, on their intervals of definition. If a function is differentiable, it must be continuous. The process of finding the derivative of a function is called **differentiation**.

Differentiation Rules

In the following formulas, u, v, and w are differentiable functions of x, and c and m are constants.

Rule 1. $\dfrac{d}{dx}(c)=0$

Rule 2. $\dfrac{d}{dx}(x)=1$

Rule 3. $\dfrac{d}{dx}(u+v+\cdots)=\dfrac{d}{dx}(u)+\dfrac{d}{dx}(v)+\cdots$

Rule 4. $\dfrac{d}{dx}(cu)=c\dfrac{d}{dx}(u)$

Rule 5. $\dfrac{d}{dx}(uv)=u\dfrac{d}{dx}(v)+v\dfrac{d}{dx}(u)$

Rule 6. $\dfrac{d}{dx}(uvw)=uv\dfrac{d}{dx}(w)+uw\dfrac{d}{dx}(v)+vw\dfrac{d}{dx}(u)$

Rule 7. $\dfrac{d}{dx}\left(\dfrac{u}{c}\right)=\dfrac{1}{c}\dfrac{d}{dx}(u),\ c\neq 0$

Rule 8. $\dfrac{d}{dx}\left(\dfrac{c}{u}\right)=c\dfrac{d}{dx}\left(\dfrac{1}{u}\right)=-\dfrac{c}{u^2}\dfrac{d}{dx}(u),\ u\neq 0$

Rule 9. $\dfrac{d}{dx}\left(\dfrac{u}{v}\right)=\dfrac{v\dfrac{d}{dx}(u)-u\dfrac{d}{dx}(v)}{v^2},\ v\neq 0$

Rule 10. $\dfrac{d}{dx}(x^m)=mx^{m-1}$

Rule 11. $\dfrac{d}{dx}(u^m)=mu^{m-1}\dfrac{d}{dx}(u)$

Example 2.3 Differentiate $y = 4 + 2x - 3x^2 - 5x^3 - 8x^4 + 9x^5$

$$\frac{dy}{dx} = 0 + 2(1) - 3(2x) - 5(3x^2) - 8(4x^3) + 9(5x^4)$$

$$= 2 - 6x - 15x^2 - 32x^3 + 45x^4$$

Example 2.4 Differentiate $y = \dfrac{3-2x}{3+2x}$.

$$y' = \frac{(3+2x)\dfrac{d}{dx}(3-2x) - (3-2x)\dfrac{d}{dx}(3+2x)}{(3+2x)^2}$$

$$= \frac{(3+2x)(-2) - (3-2x)(2)}{(3+2x)^2} = \frac{-12}{(3+2x)^2}$$

Inverse Functions

Two functions f and g such that g(f(x)) = x and f(g(y)) = y are said to be **inverse functions**. Inverse functions reverse the effect of each other. Specifically, if f(a) = b, then g(b) = a.

Example 2.5

(a) The inverse of f(x) = x + 1 is the function g(y) = y - 1.
(b) The inverse of f(x) = - x is the same function.
(c) The inverse of f(x) = $\sqrt{x}$ is the function g(y) = y^2 (defined for y ≥ 0).

(d) The inverse of f(x) = 2x - 1 is the function g(y) = y = $\dfrac{y + 1}{2}$

To obtain the inverse of a function, solve the equation y = f(x) or x in terms of y, if possible.

Not every function has an inverse function. For example, the function $f(x) = x^2$ does not possess an inverse. Since $f(1) = f(-1) = 1$, an inverse function g would have to satisfy $g(1) = 1$ and $g(1) = -1$, which is impossible. However, if we restrict the function $f(x) = x^2$ to the domain $x \geq 0$, then the function $g(y) = y^2$ would be an inverse of f which we recognize as the function $y = \pm \sqrt{x}$. The condition that a function f must satisfy to have an inverse is that f is **one-to-one**; that is, for any x_1 and x_2 in the domain of f, if $x_1 \neq x_2$, then $f(x_1) \neq f(x_2)$.

Notation: The inverse of f is denoted f^{-1}. If $y = f(x)$, we often write $x = f^{-1}(y)$. If f is differentiable, we write, as usual, dy/dx for the derivative $f'(x)$, and dx/dy for the derivative $(f^{-1})'(y)$.

If a function f has an inverse and we are given a formula for $f(x)$, then to find a formula for the inverse f^{-1}, we solve the equation $y = f(x)$ for x in terms of y. For example, given $f(x) = 5x + 2$, set $y = 5x + 2$. Then,

$$x = \frac{y-2}{5}$$

and solving for y, we obtain a formula for the inverse function:

$$f^{-1}(x) = \frac{x-2}{5}$$

The differentiation formula for finding dy/dx given dx/dy:

Rule 12. $\quad \dfrac{dy}{dx} = \dfrac{1}{(dx/dy)}$

Example 2.6 Find dy/dx, given $x = \sqrt{y} + 5$.

First method:

Solve for $y = (x - 5)^2$. Then $dy/dx = 2(x - 5)$.

Second method:

Differentiate to find

$$\frac{dx}{dy} = \frac{1}{2} y^{-1/2} = \frac{1}{2\sqrt{y}}$$

Then, by Rule 12,

$$\frac{dy}{dx} = 2\sqrt{y} = 2(x-5)$$

The Chain Rule

For two functions f and g, the function given by the formula $f(g(x))$ is called a **composite function**. If f and g are differentiable, then so is the composite function, and its derivative may be obtained by either of two procedures. The first is to compute an explicit formula for $f(g(x))$ and differentiate.

Example 2.7 If $f(x) = x^2 + 3$ and $g(x) = 2x + 1$, then

$y = f(g(x)) = (2x + 1)^2 + 3 = 4x^2 + 4x + 4$ and

$$\frac{dy}{dx} = 8x + 4 \qquad .$$

The derivative of a composite function may also be obtained with the following rule:

Rule 13. **The Chain Rule**: $D_x(f(g(x))) = f'(g(x))g'(x)$
If f is called the outer function and g is called the inner function, then $D_x(f(g(x)))$ is the product of the derivative of the outer function [evaluated at $g(x)$] and the derivative of the inner function.

Example 2.8 In Example 2.7, $f'(x) = 2x$. Therefore, $f'(g(x)) = 2g(x)$ and $g'(x) = 2$. Hence by the chain rule,

$$D_x(f(g(x))) = f'(g(x))g'(x) = 2g(x) \cdot 2 = 4g(x) = 4(2x + 1) = 8x + 4$$

You Need To Know

An alternative formulation of the chain rule is the following: Write y = f(u) and u = g(x). Then the composite function is y = f(u) = f(g(x)), and we have:

The Chain Rule: $\dfrac{dy}{dx} = \dfrac{dy}{du}\dfrac{du}{dx}$

Example 2.9 Let $y = u^3$ and $u = 4x^2 - 2x + 5$. Then the composite function $y = (4x^2 - 2x + 5)^3$ has the derivative:

$$\frac{dy}{dx} = \frac{dy}{du}\frac{du}{dx} = 3u^2(8x-2) = 3(4x^2 - 2x + 5)^2(8x-2)$$

Note: In the second formulation of the chain rule,

$$\frac{dy}{dx} = \frac{dy}{du}\frac{du}{dx},$$

the y on the left denotes the composite function of x, whereas the y on the right denotes the original function of u (what we called the outer function before).

Example 2.10 Differentiate $y = (x^2 + 4)^2(2x^3 - 1)^3$.

$$y' = (x^2+4)^2\frac{d}{dx}(2x^3-1)^3 + (2x^3-1)^3\frac{d}{dx}(x^2+4)^2$$

$$= (x^2+4)^2(3)(2x^3-1)^2\frac{d}{dx}(2x^3-1) + (2x^3-1)_3(2)(x^2+4)\frac{d}{dx}(x^2+4$$

$$= (x^2+4)^2(3)(2x^3-1)^2(6x^2) + (2x^3-1)_3(2)(x^2+4)(2x)$$

$$= 2x(x^2+4)(2x^3-1)^2(13x^3+36x-2)$$

Higher Derivatives

Let $y = f(x)$ be a differentiable function of x, and let its derivative be called the **first derivative** of the function. If the first derivative is differentiable, its derivative is called the **second derivative** of the (original) function and is denoted by one of the symbols

$$\frac{d^2y}{dx^2}, \text{ y", or } f''(x)$$

In turn, the derivative of the second derivative is called the **third derivative** of the function and is denoted by one of the symbols

$$\frac{d^3y}{dx^3}, \text{ y''', or } f'''(x)$$

and so on.

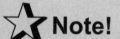

Note!

The derivative of a given order at a point can exist only when the function and all derivatives of lower order are differentiable at the point.

Example 2.11 Given

$$f(x) = \frac{2}{1-x} = 2(1-x)^{-1}, \text{ find } f^{(n)}(x).$$

$$f'(x) = 2(-1)(1-x)^{-2}(-1) = 2(1-x)^{-2} = 2(1!)(1-x)^{-2}$$
$$f''(x) = 2(1!)(-2)(1-x)^{-3}(-1) = 2(2!)(1-x)^{-3}$$
$$f'''(x) = 2(2!)(-3)(1-x)^{-4}(-1) = 2(3!)(1-x)^{-4}$$

which suggests $f^{(n)}(x) = 2(n!)(1-x)^{-(n+1)}$. This result may be established by mathematical induction by showing that if

$$f^{(k)}(x) = 2(k!)(1-x)^{-(k+1)}, \text{ then}$$

$$f^{(k+1)}(x) = -2(k!)(k+1)(1-x)^{-(k+2)}(-1) = 2[(k+1)!](1-x)^{-(k+2)}$$

Implicit Differentiation

An equation $f(x, y) = 0$, on perhaps certain restricted ranges of the variables, is said to define y **implicitly** as a function of x.

Example 2.12

(a) The equation $xy + x - 2y - 1 = 0$, with $x \neq 2$, defines the function

$$y = \frac{1-x}{x-2}$$

(b) The equation $4x^2 + 9y^2 - 36 = 0$ defines the function

$$y = \frac{2}{3}\sqrt{9-x^2}$$

when $|x| \leq 3$ and $y \geq 0$, and the function

$$y = -\frac{2}{3}\sqrt{9-x^2}$$

The derivative y' may be obtained by one of the following procedures:

- Solve, when possible, for y **explicitly** in terms of x and differentiate with respect to x. Except for very simple equations, this procedure will often prove impractical.

- Thinking of y as a function of x, differentiate both sides of the given equation with respect to x and solve the resulting relation for y'. This differentiation process is known as **implicit differentiation**.

when |x| ≤ 3 and y ≤ 0. This describes an ellipse determined by the given equation, which should be thought of as consisting of two arcs joined at the points (-3, 0) and (3, 0).

Examples 2.13
(a) Find y', given xy + y - 2y - 1 = 0.

We have

$$\left[x\frac{d}{dx}(y)+y\frac{d}{dx}(x)\right]+\frac{d}{dx}(x)-2\frac{d}{dx}(y)-\frac{d}{dx}(1)=\frac{d}{dx}(0)$$

or xy' + y + 1 - 2y' = 0; then

$$y'=\frac{1+y}{2-x}$$

(b) Find y' when x = $\sqrt{5}$, given 4x² + 9y² - 36 = 0.

We have

$$4\frac{d}{dx}(x^2)+9\frac{d}{dx}(y^2)+\frac{d}{dx}(-36)=8x+9\frac{d}{dy}(y^2)\frac{dy}{dx}=8x+18yy'=0$$

or y' = -4x/9y. When x = $\sqrt{5}$, y = ±4/3. At the point ($\sqrt{5}$, 4/3) on the upper arc of the ellipse, y' = $-\sqrt{5}/3$, and at the point ($\sqrt{5}$, - 4/3) on the lower arc, y' = $\sqrt{5}/3$.

Derivatives of Higher Order

Derivatives of higher order may be obtained in two ways:

- The first method is to differentiate implicitly the derivative of one lower order and replace y' by the relation previously found.

Example 2.14 From Example 2.13 (a),

$$y' = \frac{1+y}{2-x}$$

Then

$$\frac{d}{dx}(y') = y'' = \frac{d}{dx}\left(\frac{1+y}{2-x}\right) = \frac{(2-x)y' - (1+y)(-1)}{(2-x)^2} = \frac{(2-x)\left(\frac{1+y}{2-x}\right) + 1+y}{(2-x)^2}$$

$$= \frac{2+2y}{(2-x)^2}$$

- The second method is to differentiate implicitly both sides of the given equation as many times as is necessary to produce the required derivative and eliminate all derivatives of lower order. This procedure is recommended only when a derivative of higher order at a given point is required.

Example 2.15 Find the value of y" at the point (-1, 1) of the curve $x^2y + 3y - 4 = 0$.

We differentiate implicitly with respect to x twice, obtaining

$$(x^2y' + 2xy) + 3y' = 0$$

and

$$[(x^2y'' + 2xy') + (2xy' + 2y)] + 3y'' = 0$$

We substitute x = -1, y = 1 in the first relation to obtain y' = 1/2. Then we substitute x = -1, y = 1, y' = 1/2 in the second relation to get y" = 0.

Solved Problems

Solved Problem 2.1 Given $y = f(x) = x^2 + 5x - 8$, find Δy and $\Delta y/\Delta x$ as x changes from $x_0 = 1$ to $x_1 = x_0 + \Delta x = 1.2$.

Solution. $\Delta x = x_1 - x_o = 1.2 - 1 = 0.2$. $\Delta y = f(x_o + \Delta x) - f(x_o) = f(1.2)$ - f(1) = - 0.56 - (- 2) = 1.44. So

$$\frac{\Delta y}{\Delta x} = \frac{1.44}{0.2} = 7.2$$

Solved Problem 2.2 Differentiate

$$y = \frac{1}{x} + \frac{3}{x^2} + \frac{2}{x^3} = x^{-1} + 3x^{-2} + 2x^{-3}$$

Solution.

$$\frac{dy}{dx} = -x^{-2} + 3\left(-2x^{-3}\right) + 2\left(-3x^{-4}\right)$$
$$= -x^{-2} - 6x^{-3} - 6x^{-4}$$
$$= -\frac{1}{x^2} - \frac{6}{x^3} - \frac{6}{x^4}$$

Solved Problem 2.3 Differentiate

$$s = \left(t^2 - 3\right)^4$$

Solution.

$$\frac{ds}{dt} = 4\left(t^2 - 3\right)^3 (2t) = 8t\left(t^2 - 3\right)^3$$

Solved Problem 2.4 Differentiate

$$f(x) = x^2 + x^4 + x^6$$

Solution.

$$f'(x)=\frac{d}{dx}(x^2+x^4+x^6)=\frac{d}{dx}(x^2)+\frac{d}{dx}(x^4)+\frac{d}{dx}(x^6)$$
$$= 2x+4x^3+6x^5$$

Solved Problem 2.5 Differentiate

$$f(x)=\frac{(x^2+2)}{x^2},\ x>0$$

Solution.

$$f'(x)=\frac{(x^2)(2x)-(x^2+2)(2x)}{(x^2)^2}=\frac{2x^3-2x^3-4x}{x^4}=-\frac{4x}{x^4}$$

Solved Problem 2.6 Given $f(x) = 1 - x^3$, find $f'(-4)$ and $f'(4)$.

Solution. We first must find the derivative of $f(x)$:

$$f'(x)=-3x^2$$

Therefore,

$$f'(-4)=-3(-4)^2=-3\cdot16=-48$$

$$f'(4)=-3(4)^2=-3\cdot16=-48$$

Solved Problem 2.7 Differentiate the function:

$$f(x) = \frac{ax^2 + bx + c}{dx^2 + ex + f}$$

Solution. We must use the quotient rule to find the derivative of f(x):

$$f'(x) = \frac{(dx^2 + ex + f)(2ax + b) - (ax^2 + bx + c)(2dx + e)}{(dx^2 + ex + f)^2}$$

Solved Problem 2.8   Determine the rate of change of the area of a circle with respect to its radius, R. Also, evaluate the rate of change when R = 5.

Solution. The area of a circle is related to the radius by the function:

$$A = \pi R^2$$

Therefore, the rate of change of the area of the circle in terms of the radius, R, is

$$\frac{dA}{dR} = 2\pi R$$

which is the circumference of a circle. When R is 5,

$$\frac{dA}{dR} = 2\pi R = 2\pi \, (5) = 10\pi$$

Solved Problem 2.9   Determine the rate of change of the height, h, in terms of the radius, R, for the volume of a circular cylinder assuming a constant volume as R increases. The formula of a circular cylinder is $V = \pi R^2 h$.

Solution. Given in the problem, the volume of a circular cylinder is:

$V = \pi R^2 h$

To determine the rate of change of the cylinder volume with respect to the radius, R, we take the following derivative:

$$\frac{dV}{dR} = (\pi R^2)\frac{dh}{dR} + h\frac{d}{dR}(\pi R^2) = (\pi R^2)\frac{dh}{dR} + 2\pi Rh$$

Given that V remains constant,

$$\frac{dV}{dR} = 0 \quad \text{and thus}$$

$$\pi R^2\frac{dh}{dR} + 2\pi Rh = 0$$

Dividing through by πR yields

$$R\frac{dh}{dR} + 2h = 0$$

Solving for $\dfrac{dh}{dR}$:

$$\frac{dh}{dR} = -\frac{2h}{R}$$

Solved Problem 2.10 Determine dy/dx given:

$$y = 4u^2 + 4 \quad \text{and} \quad u = \frac{1}{x+1}$$

Solution. This is an application of the chain rule. In order to calculate dy/dx, we need to calculate dy/du and du/dx.

$$\frac{dy}{du} = 8u = \frac{8}{x+1} \; ; \; \frac{du}{dx} = -\frac{1}{(x+1)^2}$$

Therefore, using the chain rule:

$$\frac{dy}{dx} = \frac{dy}{du}\frac{du}{dx} = \left(\frac{8}{x+1}\right)\left(-\frac{1}{(x+1)^2}\right) = -\frac{8}{(x+1)^3}$$

IN THIS CHAPTER:

✔ *Tangents*
✔ *Normals*
✔ *Angle of Intersection*
✔ *Maximum and Minimum Values*
✔ *Applied Problems Involving Maxima and Minima*
✔ *Solved Problems*

Tangents

If the function $f(x)$ has a finite derivative $f'(x_o)$ at $x = x_o$, the curve $y = f(x)$ has a **tangent** at $P_o(x_o, y_o)$ whose slope is

$m = \tan \theta = f'(x_o)$

If $m = 0$, the curve has a horizontal tangent of equation $y = y_o$ at P_o, as at A, C, and E of Figure 3-1. Otherwise, the equation of the tangent is

$y - y_0 = m(x - x_0)$

If $f(x)$ is continuous at $x = x_0$ but $\lim\limits_{x \to x_0} f'(x) = \infty$, the curve has a vertical tangent given by the equation $x = x_0$, as at B and D of Figure 3-1.

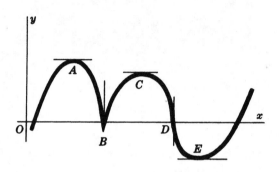

Figure 3-1

Normals

The **normal** to a curve at one of its points is the line that passes through the point and is perpendicular to the tangent at the point. Hence, if m is the slope of the tangent, then -1/m is the slope of the normal. The equation of the normal at $P_0(x_0, y_0)$ is

$x = x_0$ if the tangent is horizontal

$y = y_0$ if the tangent is vertical

$y - y_0 = -\dfrac{1}{m}(x - x_0)$ otherwise

Example 3.1 Find the equations of the tangent and normal to $y = x^3 - 2x^2 + 4$ at $(2, 4)$.

$f'(x) = 3x^2 - 4x$; hence the slope of the tangent at $(2, 4)$ is $m = f'(2) = 4$.

The equation of the tangent is $y - 4 = 4(x - 2)$ or $y = 4x - 4$.
The equation of the normal is

$$y - 4 = -\frac{1}{4}(x-2) \text{ or } x + 4y = 18$$

Example 3.2 Find the equation of the line containing the point $(2, -2)$, which is tangent to the hyperbola $x^2 - y^2 = 16$.

Let $P_o(x_o, y_o)$ be the point of tangency. Then, P_o is on the hyperbola, so

$$x_o^2 - y_o^2 = 16 \qquad (3\text{-}1)$$

Also,

$$\frac{dy}{dx} = \frac{x}{y}$$

Hence, at (x_o, y_o), the slope of the line joining P_o and $(2, -2)$ is

$$m = \frac{x_o}{y_o} = \frac{y_o - (-2)}{x_o - 2}$$

Then,

$$2x_o + 2y_o = x_o^2 - y_o^2 = 16 \text{ or } x_o + y_o = 8 \qquad (3\text{-}2)$$

The point of tangency is the simultaneous solution $(5, 3)$ of Eqs. (3-1) and (3-2). Thus, the equation of the tangent is

$$y - 3 = \frac{5}{3}(x-5)$$

or

$$5x - 3y = 16$$

Angle of Intersection

The **angle of intersection** of two curves is defined as the angle between the tangents to the curves at their point of intersection.

To determine the angles of intersection of two curves:

1. Solve the equations simultaneously to find the points of intersection.
2. Find the slopes m_1 and m_2 of the tangents to the two curves at each point of intersection.
3. If $\mathbf{m_1 = m_2}$ the angle of intersection is $\phi = 0°$. If $\mathbf{m_1 = -1/m_2}$ the angle of intersection is $\phi = 90°$.

Otherwise, it can be found from

$$\tan \varphi = \frac{m_1 - m_2}{1 + m_1 m_2}$$

where ϕ is the acute angle of intersection when $\tan \phi > 0$, and $180° - \phi$ is the acute angle of intersection when $\tan \phi < 0$.

Example 3.3 A cable of a certain suspension bridge is attached to supporting pillars 250 feet (ft) apart. If it hangs in the form of a parabola with the lowest point 50 ft below the point of suspension, find the angle between the cable and the pillar.

Take the origin at the vertex of the parabola, as in Figure 3-2. The equation of the parabola is $y = (2/625)x^2$ and $y'=4x/625$. At (125, 50), m = 4(125)/625 = 0.8000 and $\theta = 38°40'$. Hence the required angle is $\phi = 90° - \theta = 51°20'$.

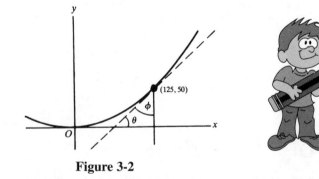

Figure 3-2

Maximum and Minimum Values

Increasing and Decreasing Functions

A function $f(x)$ is said to be **increasing** on an open interval if $u < v$ implies $f(u) < f(v)$ for all u and v in the interval. A function $f(x)$ is said to be increasing at $x = x_0$ if $f(x)$ is increasing on an open interval containing x_0. Similarly, $f(x)$ is **decreasing** on an open interval if $u < v$ implies $f(u) > f(v)$ for all u and v in the interval, and $f(x)$ is decreasing at $x = x_0$ if $f(x)$ is decreasing on an open interval containing x_0.

If $f'(x_0) > 0$, then it can be shown that $f(x)$ is an increasing function at $x = x_0$; similarly, if $f'(x_0) < 0$, then $f(x)$ is a decreasing function at $x = x_0$. If $f'(x_0) = 0$, then $f(x)$ is said to be **stationary** at $x = x_0$.

In Figure 3-3, the curve $y = f(x)$ is rising (the function is increasing) on the intervals $a < x < r$ and $t < x < u$; the curve is falling (the function is decreasing) on the interval $r < x < t$. The function is stationary at $x = r$, $x = s$, and $x = t$; the curve has horizontal tangents at points R, S, and T. The values of x (that is, r, s, and t), for which the function $f(x)$ is stationary [that is, for $f'(x) = 0$] are **critical values** for the function, and the corresponding points (R, S, and T) of the graph are called **critical points** of the curve.

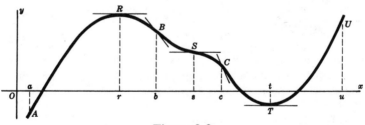

Figure 3-3

Relative Maximum and Minimum Values of a Function

A function $f(x)$ is said to have a **relative maximum** at $x = x_o$ if $f(x_o) \geq f(x)$ for all x in some open interval containing x_o, that is, if the value of $f(x_o)$ is greater than or equal to the values of $f(x)$ at all nearby points. A function $f(x)$ is said to have a **relative minimum** at $x = x_o$ if $f(x_o) \leq f(x)$ for all x in some open interval containing x_o, that is, if the value of $f(x_o)$ is less than or equal to the values of $f(x)$ at all nearby points.

In Figure 3-3, $R(r, f(r))$ is a relative maximum point of the curve since $f(r) > f(x)$ on any sufficiently small neighborhood $0 < |x - r| < \delta$. We say that $y = f(x)$ has a relative maximum value $(= f(r))$ when $x = r$. In the same figure, $T(t, f(t))$ is a relative minimum point of the curve since $f(t) < f(x)$ on any sufficiently small neighborhood $0 < |x - t| < \delta$. We say that $y = f(x)$ has a relative minimum value $(= f(t))$ when $x = t$. Note that R joins an arc AR which is rising $(f'(x) > 0)$ and an arc RB which is falling $(f'(x) < 0)$, while T joins an arc CT which is falling $(f'(x) < 0)$ and an arc TU which is rising $(f'(x) > 0)$. At S, two arcs BS and SC, both of which are falling, are joined; S is neither a relative maximum point nor a relative minimum point of the curve.

If $f(x)$ is differentiable on $a \leq x \leq b$ and if $f(x)$ has a relative maximum (minimum) value at $x = x_o$, where $a < x_o < b$, then $f'(x_o) = 0$.

First Derivative Test

The following steps can be used to find the relative maximum (or minimum) values (hereafter called simply maximum [or minimum] values) of a function $f(x)$ that, together with its first derivative, is continuous.

1. Solve $f'(x) = 0$ for the critical values.
2. Locate the critical values on the x axis, thereby establishing a number of intervals.
3. Determine the sign of $f'(x)$ on each interval.
4. Let x increase through each critical value $x = x_o$; then:

    (a) $f(x)$ has a maximum value $f(x_o)$ if $f'(x)$ changes from + to - (Figure 3-4).

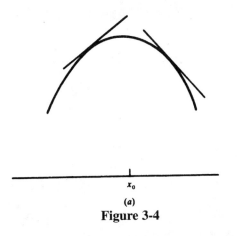

(a)

Figure 3-4

(b) f(x) has a minimum value $f(x_0)$ if $f'(x)$ changes from - to + (Figure 3-5).

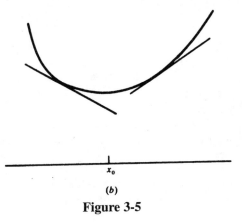

(b)

Figure 3-5

(c) f(x) has neither a maximum nor a minimum value at $x = x_0$ if $f'(x)$ does not change sign (Figure 3-6).

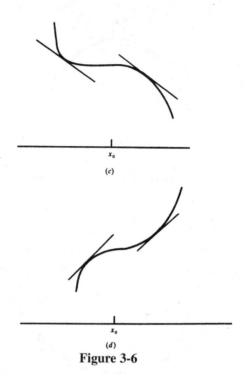

(c)

(d)

Figure 3-6

Example 3.4 Given

$$y = \frac{1}{3}x^3 + \frac{1}{2}x^2 - 6x + 8$$

find (a) the critical points; (b) the intervals on which y is increasing and decreasing; and (c) the maximum and minimum values of y.

(a) $y' = x^2 + x - 6 = (x + 3)(x - 2)$. Setting $y' = 0$ gives the critical values $x = -3$ and 2. The critical points are $(-3, 43/2)$ and $(2, 2/3)$.

(b) When y' is positive, y increases; when y' is negative, y decreases.
When $x < -3$, say $x = -4$, $y' = (-)(-) = +$, and y is increasing
When $-3 < x < 2$, say $x = 0$, $y' = (+)(-) = -$, and y is decreasing
When $x > 2$, say $x = 3$, $y' = (+)(+) = +$, and y is increasing

These results are illustrated by Figure 3-7:

Remember

A function f(x) may have a maximum or minimum value $f(x_o)$ although $f'(x_o)$ does not exist. The values $x = x_o$ for which f(x) is defined but f'(x) does not exist will also be called critical values for the function. *They, together with the values for which $f'(x) = 0$, are to be used as the critical values in the first-derivative test.*

$x < -3$	$x = -3$	$-3 < x < 2$	$x = 2$	$x > 2$
$y' = +$		$y' = -$		$y' = +$
y increases		y decreases		y increases

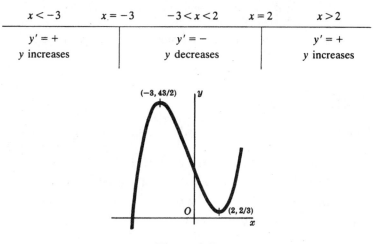

Figure 3-7

(c) We test the critical values x = - 3 and 2 for maxima and minima.

As x increases through - 3, y' changes sign from + to -; hence at x = - 3, y has a maximum value 43/2.

As x increases through 2, y' changes sign from - to +; hence at x = 2, y has a minimum value 2/3.

Example 3.5 Examine $y = |x|$ for maximum and minimum values.

The function is everywhere defined and has a derivative for all x except $x = 0$. Thus, $x = 0$ is a critical value. For $x < 0$, $f'(x) = -1$; for $x > 0$, $f'(x) = +1$. The function has a minimum $(= 0)$ when $x = 0$.

Concavity

An arc of a curve $y = f(x)$ is called **concave upward** if, at each of its points, the arc lies above the tangent at that point. As x increases, $f'(x)$ either is of the same sign and increasing (as on the interval $b < x < s$ of Figure 3-3) or changes sign from negative to positive (as on the interval $c < x < u$). In either case, the slope $f'(x)$ is increasing hence $f''(x) > 0$.

An arc of a curve $y = f(x)$ is called **concave downward** if, at each of its points, the arc lies below the tangent at that point. As x increases, $f'(x)$ either is of the same sign and decreasing (as on the interval $s < x < c$ of Figure 3-3) or changes sign from positive to negative (as on the interval $a < x < b$). In either case, the slope $f'(x)$ is decreasing and $f''(x) < 0$.

Point of Inflection

A **point of inflection** is a point at which a curve changes from concave upward to concave downward, or vice versa. In Figure 3-3, the points of inflection are B, S, and C. A curve $y = f(x)$ has one of its points $x = x_0$ as an inflection point if $f''(x_0) = 0$ or is not defined *and* $f''(x)$ changes sign between points $x < x_0$ and $x > x_0$. (The latter condition may be replaced by $f'''(x_0) \neq 0$ when $f'''(x_0)$ exists.)

Example 3.6 Examine $y = x^4 - 6x + 2$ for concavity and points of inflection.

The graph of the function is shown in Figure 3-8.

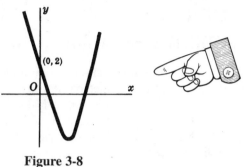

Figure 3-8

We have $y'' = 12x^2$. The possible point of inflection is at $x = 0$. On the intervals $x < 0$ and $x > 0$, $y'' = +$; hence the arcs on both sides of $x = 0$ are concave upward. Therefore, the point $(0, 2)$ is not a point of inflection.

Second-Derivative Test

There is a second test for maxima and minima:

1. Solve $f'(x_o) = 0$ and determine where $f'(x_o)$ does not exist for the critical values.
2. For each critical value $x = x_o$:

   $f(x)$ has a maximum value $f(x_o)$ if $f''(x_o) < 0$ (Figure 3-4)

   $f(x)$ has a minimum value $f(x_o)$ if $f''(x_o) > 0$ (Figure 3-5)

   The test fails if $f''(x_o) = 0$ or is not defined (Figure 3-6). In this case, the first derivative test must be used.

Example 3.7 Examine $f(x) = x(12 - 2x)^2$ for maxima and minima using the second-derivative method.

Here $f'(x) = 12(x^2 - 8x + 12) = 12(x - 2)(x - 6)$. Hence, the critical values are $x = 2$ and 6. Also, $f''(x) = 12(2x - 8) = 24(x - 4)$. Because $f''(2) < 0$, $f(x)$ has a maximum value $(= 128)$ at $x = 2$. Because $f''(6) > 0$, $f(x)$ has a minimum value $(= 0)$ at $x = 6$.

Applied Problems Involving Maxima and Minima

To determine **absolute maxima and minima** over a closed interval [a, b], we use the following method (in lieu of either the first- or second-derivative tests): First, identify all critical values, c. Then, determine the value of the function y = f(x) at each of the endpoints f(a) and f(b) and at each critical value f(c). Finally, compare these to obtain the maximum and minimum values.

Example 3.8 Divide the number 120 into two parts such that the product P of one part and the square of the other is a maximum.

Let x be one part, and 120 - x the other part. Then, P = (120 - x)x², and 0 ≤ x ≤ 120. Since dP/dx = 3x(80 - x), the critical values are x = 0 and x = 80. Now P(0) = 0, P(80) = 256,000, and P(120) = 0; hence the maximum value of P occurs when x = 80. The required parts are 80 and 40.

Example 3.9 A cylindrical container with a circular base is to hold 64 cubic inches (in³). Find its dimensions so that the amount (surface area) of metal required is a minimum when the container is (a) an open cup and (b) a closed can.

Let r and h be, respectively, the radius of the base and the height in inches, A the amount of metal, and V the volume of the container.

(a) Here $V = \pi r^2 h = 64$ in³, and $A = 2\pi rh + \pi r^2$. To express A as a function of one variable, we solve for h in the first relation (because it is easier) and substitute in the second, obtaining

$$A = 2\pi r \frac{64}{\pi r^2} + \pi r^2 = \frac{128}{r} + \pi r^2 \text{ and } \frac{dA}{dr} = -\frac{128}{r^2} + 2\pi r = \frac{2(\pi r^3 - 64)}{r^2}$$

and the critical value is $r = 4/\sqrt[3]{\pi}$. Then h = 64/πr² = $4/\sqrt[3]{\pi}$. Thus,

$r = h = 4/\sqrt[3]{\pi}$ in.

Now dA/dr > 0 to the right of the critical value, and dA/dr < 0 to the left of the critical value. So, by the first-derivative test, we have a relative minimum. Since there is no other critical value, that relative minimum is an absolute minimum.

(b) Here again $V = \pi r^2 h = 64$ in^3, but $A = 2\pi rh + 2\pi r^2 = 2\pi r(64/\pi r^2) + 2\pi r^2 = 128/r + 2\pi r^2$. Hence

$$\frac{dA}{dr} = -\frac{128}{r^2} + 4\pi r = \frac{4(\pi r^3 - 32)}{r^2}$$

and the critical value is $r = 2\sqrt[3]{4/\pi}$. Then $h = 64/\pi r^2 = 2\sqrt[3]{4/\pi}$. Thus, $h = 2r = 4\sqrt[3]{4/\pi}$ in. That we have found an absolute minimum can be shown as in part (a).

Example 3.10 A man in a rowboat at P in Figure 3-9, 5 miles (mi) from the nearest point A on a straight shore, wishes to reach a point B, 6 mi from A along the shore, in the shortest time. Where should he land if he can row 2 miles per hour (mi/h) and walk 4 mi/h?

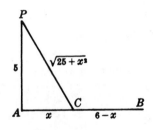

Figure 3-9

Let C be the point between A and B at which the man lands, and let AC = x. The distance rowed is $PC = \sqrt{25 + x^2}$, and the rowing time required is

$$t_1 = \frac{\text{distance}}{\text{speed}} = \frac{\sqrt{25 + x^2}}{2}$$

The distance walked is CB = 6 - x, and the walking time required is

$t_2 = (6 - x)/4$. Hence, the total time required is

$$t = t_1 + t_2 = \frac{1}{2}\sqrt{25+x^2} + \frac{1}{4}(6-x) \quad \text{and}$$

$$\frac{dt}{dx} = \frac{x}{2\sqrt{25+x^2}} - \frac{1}{4} = \frac{2x - \sqrt{25+x^2}}{4\sqrt{25+x^2}}$$

The critical value, obtained from $2x - \sqrt{25+x^2} = 0$, is

$$x = \frac{5}{3}\sqrt{3} \sim 2.89$$

Thus, he should land at a point 2.89 mi from A toward B.

Solved Problems

Solved Problem 3.1 Show that the curve $y = x^3 - 8$ has no maximum or minimum value.

Solution. Setting $y' = 3x^2 = 0$ gives the critical value $x = 0$. But $y' > 0$ when $x < 0$ and when $x > 0$. Hence y has no maximum or minimum value. The curve has a point of inflection at $x = 0$.

Solved Problem 3.2 Examine $y = 3x^4 - 10x^3 - 12x^2 + 12x - 7$ for concavity and points of inflection.

Solution. We have

$y' = 12x^3 - 30x^2 - 24x + 12$
$y'' + 36x^2 - 60x - 24 = 12(3x + 1)(x - 2)$

Set $y''=0$ and solve to obtain the possible points of inflection $x = -\frac{1}{3}$ and 2. Then

When $x < -\dfrac{1}{3}$, $\qquad$ $y'' = +$, and the arc is concave upward.

When $-\dfrac{1}{3} < x < 2$, $\qquad$ $y'' = -$, and the arc is concave downward.

When $x > 2$, $\qquad$ $y'' = +$, and the arc is concave upward.

The points of inflection are $\left(-\dfrac{1}{3}, -\dfrac{322}{27}\right)$ and $(2, -63)$, since y'' changes sign at $x = -\dfrac{1}{3}$ and $x = 2$ (see Figure SP3-1).

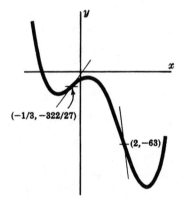

Figure SP3-1

Solved Problem 3.3 Examine $y = x^2 + 250/x$ for maxima and minima using the second-derivative method.

Solution. Here

$$y' = 2x - \frac{250}{x^2} = \frac{2(x^3 - 125)}{x^2}$$, so the critical value is x = 5.

Also, $y'' = 2 + \frac{500}{x^3}$. Because y" > 0 at x = 5, y has a minimum value (= 75) at x = 5.

Solved Problem 3.4 Using 200 feet of wire, Alexandra would like to construct a rectangular garden consisting of three sides with the fourth side against a wall of the house. What are the dimensions of the garden that will yield the maximum possible area?

Solution. We first begin by defining

x = length of the garden side perpendicular to the house

y = length of the garden side parallel to the house

Given the total amount of fencing wire is 200 feet, then

$$2x + y = 200 \tag{1}$$

Also, the area of the rectangular garden is

$$A = x \, y \tag{2}$$

Solving Eq. (1) for y in terms of x

$$y = 200 - 2x \tag{3}$$

Substituting Eq. (3) into Eq. (2)

$$A = x \, (200 - 2x) \tag{4}$$

defined over the interval $0 \le x \le 100$. Given this, we seek to determine the maximum of $A = f(x) = x (200 - 2x)$ for x in the interval [0, 100]. In this case,

$A = f(x) = 200x - 2x^2$

We take the derivative of A:

$f'(x)=200 - 4x$

Setting $f'(x) = 0$ -

$200 - 4x = 0$

$4x = 200$

$x = 50$ feet

Substituting this value into Eq. (3),

$y = 200 - 2(50) = 200 - 100 = 100$ feet

Thus, the dimensions of the garden that yield the maximum possible area are $x = 50$ feet and $y = 100$ feet.

Solved Problem 3.5 Given a square piece of cardboard with sides equal to 16 inches, Laura would like to construct a box by cutting out four squares, one from each corner. What is the size of the square that should be cut out in order to maximize the volume of the box?

Solution. We begin by defining the length of the square to be cut from each corner of the cardboard as x. Then, each side of the cardboard square is defined as:

Length $= 16 - 2x$

Therefore, the volume of the cardboard box can be determined as:

$$\text{Volume} = V(x) \quad = (\text{length})(\text{width})(\text{height})$$
$$= (16 - 2x)(16 - 2x)(x)$$
$$= 4x^3 - 64x^2 + 256x$$

The value of x that maximizes the volume fo the box is in the interval [0, 8] and can occur at either 0, 8 or at some critical number which satisfies the calculation V'(x) = 0. The values of 0 and 8 do not make sense as far as possibilities so therefore we must determine the critical values.

$$V'(x) = 12x^2 - 128x + 256$$
$$= 4(3x^2 - 32x + 64)$$
$$= 4(3x - 8)(x - 8)$$

The equation

$$V'(x) = 4(3x - 8)(x - 8) = 0$$

has two roots:

$$x = \frac{8}{3} \quad \text{and} \quad x = 8$$

Since x = 8 was eliminated previously, the value of x that yields the maximum volume is

$$x = \frac{8}{3}$$

and is equal to

$$V\left(\frac{8}{3}\right) = 4\left(\frac{8}{3}\right)^3 - 64\left(\frac{8}{3}\right)^2 + 256\left(\frac{8}{3}\right) = 455.2 \text{ cubic inches}$$

Chapter 4
DIFFERENTIATION OF SPECIAL FUNCTIONS

Differentiation of Trigonometric Functions

Radian Measure

Let s denote the length of an arc AB intercepted by the central angle AOB on a circle of radius r, and let S denote the area of the sector AOB (see Figure 4-1).

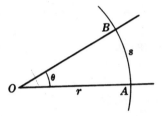

Figure 4-1

(If s is 1/360 of the circumference, then angle AOB has measure 1°; if s = r, angle AOB has measure 1 radian (rad). And, since a full circle has a circumference of 2π rad, we can write 1 rad = 180/π degrees and 1° = π/180 rad. Thus 0° = 0 rad; 30° = π/6 rad; 45° = π/4 rad; 180° = π rad; and 360° = 2π rad.)

Suppose AOB is measured as α degrees; then we formulate the arc length and area of the sector as:

$$s = \frac{\pi}{180}\alpha r \quad \text{and} \quad S = \frac{\pi}{360}\alpha r^2$$

(4.1)

Suppose next that AOB is measured as θ rad; then

$$s = \theta r \quad \text{and} \quad S = \frac{1}{2}\theta r^2$$

(4.2)

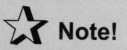 **Note!**

A comparison of Eqs. (4.1) and (4.2) will make clear one of the advantages of radian measure. In particular, it is a unitless measurement, i.e., a real number.

Trigonometric Functions

Let θ be any real number. Construct the angle whose measure is θ radians with vertex at the origin of a rectangular coordinate system, and initial side along the positive x axis (see Figure 4-2).

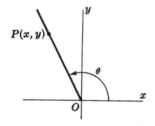

Figure 4-2

Take P(x, y) on the terminal side of the angle a unit distance from O; then we define the functions sin θ = y and cos θ = x. The domain of definition of both sin θ and cos θ is the set of real numbers; the range of sin θ is - 1 ≤ y ≤ 1, and the range of cos θ is - 1 ≤ x ≤ 1. Recall that, if θ is an acute angle of a right triangle ABC (see Figure 4-3), then

$$\sin \theta = \frac{\text{opposite side}}{\text{hypotenuse}} = \frac{BC}{AB}$$

$$\cos \theta = \frac{\text{adjacent side}}{\text{hypotenuse}} = \frac{AC}{AB}$$

$$\tan \theta = \frac{\text{opposite side}}{\text{adjacent side}} = \frac{BC}{AC}$$

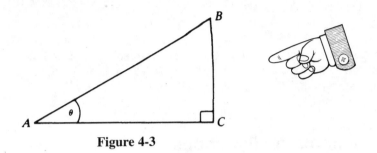

Figure 4-3

The slope m of a nonvertical line is equal to tan α, where α is the coun-
terclockwise angle from the positive x axis to the line. Table 4-1 lists
some standard trigonometric identities, and Table 4-2 contains some
useful values of the trigonometric functions.

$$\sin^2 \theta + \cos^2 \theta = 1$$

$$\sin(-\theta) = -\sin\theta, \ \cos(-\theta) = \cos\theta$$

$$\sin(\alpha + \beta) = \sin\alpha\cos\beta + \cos\alpha\sin\beta$$

$$\sin(\alpha - \beta) = \sin\alpha\cos\beta - \cos\alpha\sin\beta$$

$$\cos(\alpha + \beta) = \cos\alpha\cos\beta - \sin\alpha\sin\beta$$

$$\cos(\alpha - \beta) = \cos\alpha\cos\beta + \sin\alpha\sin\beta$$

$$\sin 2\alpha = 2\sin\alpha\cos\alpha$$

$$\cos 2\alpha = \cos^2\alpha - \sin^2\alpha = 1 - 2\sin^2\alpha = 2\cos^2\alpha - 1$$

$$\sin(\alpha + 2\pi) = \sin\alpha, \ \cos(\alpha + 2\pi) = \cos\alpha$$

$$\sin(\alpha + \pi) = -\sin\alpha, \ \cos(\alpha + \pi) = -\cos\alpha, \ \tan(\alpha + \pi) = \tan\alpha$$

$$\sin\left(\frac{\pi}{2} - \alpha\right) = \cos\alpha, \ \cos\left(\frac{\pi}{2} - \alpha\right) = \sin\alpha$$

$$\sin(\pi - \alpha) = \sin\alpha, \ \cos(\pi - \alpha) = -\cos\alpha$$

$$\sec^2\alpha = 1 + \tan^2\alpha$$

$$\tan(\alpha + \beta) = \frac{\tan\alpha + \tan\beta}{1 - \tan\alpha\tan\beta}$$

$$\tan(\alpha - \beta) = \frac{\tan\alpha - \tan\beta}{1 + \tan\alpha\tan\beta}$$

Table 4-1

x	$\sin x$	$\cos x$	$\tan x$
0	0	1	0
$\pi/6$	$1/2$	$\sqrt{3}/2$	$\sqrt{3}/3$
$\pi/4$	$\sqrt{2}/2$	$\sqrt{2}/2$	1
$\pi/3$	$\sqrt{3}/2$	$1/2$	$\sqrt{3}$
$\pi/2$	1	0	∞
π	0	-1	0
$3\pi/2$	-1	0	∞

Table 4-2

Differentiation Formulas

Now, we can define the trigonometric functions in terms of the usual variable x (rather than denoting the angle argument θ). Then, we can derive the following formulas:

Rule 14. $\dfrac{d}{dx}(\sin x)=\cos x$ Rule 15. $\dfrac{d}{dx}(\cos x)=-\sin x$

Rule 16. $\dfrac{d}{dx}(\tan x)=\sec^2 x$ Rule 17. $\dfrac{d}{dx}(\cot x)=-\csc^2 x$

Rule 18. $\dfrac{d}{dx}(\sec x)=\sec x \tan x$ Rule 19. $\dfrac{d}{dx}(\csc x)=-\csc x \cot x$

Example 4.1 Find the first derivative of y = sin 3x + cos 2x.

$$y'=\cos 3x\frac{d}{dx}(3x)-\sin 2x\frac{d}{dx}(2x)=3\cos 3x-2\sin 2x$$

Example 4.2 Find the first derivative of $f(x)=\dfrac{\cos x}{x}$.

$$f'(x) = \frac{x\frac{d}{dx}(\cos x) - \cos x\frac{d}{dx}(x)}{x^2} = \frac{-x\sin x - \cos x}{x^2}$$

Differentiation of Inverse Trigonometric Functions

The Inverse Trigonometric Functions

If x = sin y, the inverse function is written y = arcsin x. (An alternative notation is y = sin^{-1} x.) The domain of arcsin x is - 1 ≤ x ≤ 1, which is the range of sin y. The range of arcsin x is the set of real numbers, which is the domain of sin y. The domain and range of the remaining inverse trigonometric functions may be established in a similar manner.

The inverse trigonometric functions are multivalued. In order that there be agreement on separating the graph into single-valued arcs, we define in Table 4-3 one such arc (called the **principal branch**) for each function. In Figure 4-4, the principal branches are indicated by a thicker curve.

Function	Principal Branch
$y = \arcsin x$	$-\frac{1}{2}\pi \leqq y \leqq \frac{1}{2}\pi$
$y = \arccos x$	$0 \leqq y \leqq \pi$
$y = \arctan x$	$-\frac{1}{2}\pi < y < \frac{1}{2}\pi$
$y = \text{arccot } x$	$0 < y < \pi$
$y = \text{arcsec } x$	$-\pi \leqq y < -\frac{1}{2}\pi,\ 0 \leqq y < \frac{1}{2}\pi$
$y = \text{arccsc } x$	$-\pi < y \leqq -\frac{1}{2}\pi,\ 0 < y \leqq \frac{1}{2}\pi$

Table 4-3

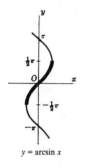

$y = \arcsin x$

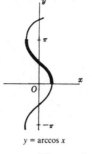

$y = \arccos x$

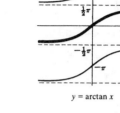

$y = \arctan x$

Figure 4-4

Differentiation Formulas

Rule 20. $\dfrac{d}{dx}(\arcsin x) = \dfrac{1}{\sqrt{1-x^2}}$

Rule 21. $\dfrac{d}{dx}(\arccos x) = -\dfrac{1}{\sqrt{1-x^2}}$

Rule 22. $\dfrac{d}{dx}(\arctan x) = \dfrac{1}{1+x^2}$

Rule 23. $\dfrac{d}{dx}(\text{arccot } x) = -\dfrac{1}{1+x^2}$

Rule 24. $\dfrac{d}{dx}(\text{arcsec } x) = \dfrac{1}{x\sqrt{x^2-1}}$

Rule 25. $\dfrac{d}{dx}(\text{arccsc } x) = -\dfrac{1}{x\sqrt{x^2-1}}$

Example 4.3 Find the first derivative of $y = \arctan 3x^2$.

$$\frac{dy}{dx} = \left[\frac{1}{1+\left(3x^2\right)^2} \right] \frac{d}{dx}\left(3x^2\right) = \frac{6x}{1+9x^4}$$

Example 4.4 Find the first derivative of $f(x) = x\sqrt{a^2-x^2} + a^2\arcsin\dfrac{x}{a}$.

$$f'(x) = x\left[\frac{1}{2}\left(a^2-x^2\right)^{-1/2}\left(-2x\right) \right] + \left(a^2-x^2\right)^{1/2} + a^2\frac{1}{\sqrt{1-\left(\dfrac{x}{a}\right)^2}}\frac{1}{a}$$

$$= 2\sqrt{a^2-x^2}$$

Differentiation of Exponential and Logarithmic Functions

Define the number e by the equation

$$e = \lim_{h\to+\infty}\left(1+\frac{1}{h}\right)^h$$

Then e also can be represented by $\lim_{k\to 0}(1+k)^{1/k}$. In addition, it can also be shown that

$$e = 1+1+\frac{1}{2!}+\frac{1}{3!}+\cdots+\frac{1}{n!}+\cdots = 2.71828$$

Called the "natural" number or Euler's number, e will serve as a base for the natural logarithm function.

Logarithmic Functions

Assume $a > 0$ and $a \neq 1$. If $a^y = x$, then define $y = \log_a x$. That is, $x = a^y$ and $y = \log_a x$ are inverse functions.

Let $\ln x = \log_e x$. Then $\ln x$ is called the **natural logarithm** of x. See also Figure 4-5. The domain of $\log_a x$ is $x > 0$; the range is the set of real numbers.

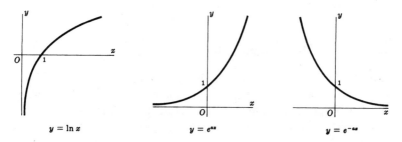

$y = \ln x$ $y = e^{ax}$ $y = e^{-as}$

Figure 4-5

Differentiation Formulas

Rule 26. $\dfrac{d}{dx}\left(\log_a x\right)=\dfrac{1}{x}\log_a e$, $a > 0$, $a\neq 1$

$$=\frac{1}{x \ln a}$$

Rule 27. $\dfrac{d}{dx}\left(\ln x\right)=\dfrac{1}{x}$

Rule 28. $\dfrac{d}{dx}\left(a^x\right)=a^x \ln a$, $a > 0$

Rule 29. $\dfrac{d}{dx}\left(e^x\right)=e^x$

Example 4.5 Find the first derivative of $y = \log_a(3x^2 - 5)$.

$$\frac{dy}{dx}=\frac{1}{3x^2-5}\left(\log_a e\right)\frac{d}{dx}\left(3x^2-5\right)=\frac{6x}{3x^2-5}\log_a e=\frac{6x}{\left(3x^2-5\right)\ln a}$$

Example 4.6 Find the first derivative of y = ln sin 3x.

$$y' = \frac{1}{\sin 3x}\frac{d}{dx}(\sin 3x) = 3\frac{\cos 3x}{\sin 3x} = 3\cot 3x$$

Logarithmic Differentiation

If a differentiable function y = f(x) is the product and/or quotient of several factors, the process of differentiation may be simplified by taking the natural logarithm of the function before differentiation since

$$\frac{d}{dx}(\ln y) = \frac{1}{y}\frac{d}{dx}(y)$$

This amounts to using the formula

Rule 30. $\dfrac{d}{dx}(y) = y\dfrac{d}{dx}(\ln\ y)$

Example 4.7 Use logarithmic differentiation to find the first derivative, given the function y = (x² + 2)³(1 + x³)⁴.

$$\ln y = \ln(x^2+2)^3(1-x^3)^4 = 3\ln(x^2+2)+4\ln(1-x^3)$$

$$y' = y\frac{d}{dx}[3\ln(x^2+2)+4\ln(1-x^3)] = (x^2+2)^3(1-x^3)^4\left(\frac{6x}{x^2+2}-\frac{12x^2}{1-x^3}\right)$$

$$= 6x(x^2+2)^2(1-x^3)^3(1-4x-3x^3)$$

Differentiation of Hyperbolic Functions

Definitions of Hyperbolic Functions

For any real number, x, except where noted, the hyperbolic functions are defined as:

$$\sinh x = \frac{e^x - e^{-x}}{2}$$

$$\cosh x = \frac{e^x + e^{-x}}{2}$$

$$\tanh x = \frac{\sinh x}{\cosh x} = \frac{e^x - e^{-x}}{e^x + e^{-x}}$$

$$\coth x = \frac{1}{\tanh x} = \frac{e^x + e^{-x}}{e^x - e^{-x}}, \; x \neq 0$$

$$\operatorname{sech} x = \frac{1}{\cosh x} = \frac{2}{e^x + e^{-x}}$$

$$\operatorname{csch} x = \frac{1}{\sinh x} = \frac{2}{e^x - e^{-x}}, \; x \neq 0$$

Differentiation Formulas

Rule 31. $\dfrac{d}{dx}(\sinh x) = \cosh x$

Rule 32. $\dfrac{d}{dx}(\cosh x) = \sinh x$

Rule 33. $\dfrac{d}{dx}(\tanh x) = \operatorname{sech}^2 x$

Rule 34. $\dfrac{d}{dx}(\coth x) = -\operatorname{csch}^2 x$

Rule 35. $\dfrac{d}{dx}(\operatorname{sech} x) = -\operatorname{sech} x \tanh x$

Rule 36. $\dfrac{d}{dx}(\operatorname{csch} x) = -\operatorname{csch} x \coth x$

Example 4.8 Find dy/dx given the function: y = sinh 3x.

$$\frac{dy}{dx} = \cosh 3x \frac{d}{dx}(3x) = 3 \cosh 3x$$

Example 4.9 Find dy/dx given the function: $y = \coth \dfrac{1}{x}$

$$\frac{dy}{dx} = -\operatorname{csch}^2 \frac{1}{x} \frac{d}{dx}\left(\frac{1}{x}\right) = \frac{1}{x^2} \operatorname{csch}^2 \frac{1}{x}$$

Differentiation of Inverse Hyperbolic Functions

Definitions of Inverse Hyperbolic Functions

$$\sinh^{-1}x = \ln\left(x + \sqrt{1+x^2}\right) \text{ for all } x \qquad \coth^{-1}x = \frac{1}{2}\ln\frac{x+1}{x-1}, \ x^2 > 1$$

$$\cosh^{-1}x = \ln\left(x + \sqrt{x^2-1}\right), \ x \geq 1 \qquad \operatorname{sech}^{-1}x = \ln\frac{1+\sqrt{1-x^2}}{x}, \ 0 < x \leq 1$$

$$\tanh^{-1}x = \frac{1}{2}\ln\frac{1+x}{1-x}, \ x^2 < 1 \qquad \operatorname{csch}^{-1}x = \ln\left(\frac{1}{x} + \frac{\sqrt{1+x^2}}{|x|}\right), \ x \neq 0$$

Differentiation Formulas

Rule 37. $\dfrac{d}{dx}\left(\sinh^{-1}x\right) = \dfrac{1}{\sqrt{1+x^2}}$

Rule 38. $\dfrac{d}{dx}\left(\cosh^{-1}x\right) = \dfrac{1}{\sqrt{x^2-1}}, \ x > 1$

Rule 39. $\dfrac{d}{dx}\left(\tanh^{-1}x\right) = \dfrac{1}{1-x^2}, \ x^2 < 1$

Rule 40. $\dfrac{d}{dx}\left(\coth^{-1}x\right) = \dfrac{1}{1-x^2}, \ x^2 > 1$

Rule 41. $\dfrac{d}{dx}\left(\operatorname{sech}^{-1}x\right) = \dfrac{-1}{x\sqrt{1-x^2}}, \ 0 < x < 1$

Rule 42. $\dfrac{d}{dx}\left(\operatorname{csch}^{-1}x\right) = \dfrac{-1}{|x|\sqrt{1+x^2}}, \ x \neq 0$

Example 4.10 Derive $\dfrac{d}{dx}\left(\sinh^{-1}x\right) = \dfrac{1}{\sqrt{1+x^2}}$

Let $y = \sinh^{-1} x$. Then $\sinh y = x$ and differentiation yields

$$\cosh y \, \frac{dy}{dx} = 1$$

so,

$$\frac{dy}{dx} = \frac{1}{\cosh y} = \frac{1}{\sqrt{1+\sinh^2 y}} = \frac{1}{\sqrt{1+x^2}}$$

Example 4.11 Find dy/dx given the function

$$y = \cosh^{-1} e^x$$

$$\frac{dy}{dx} = \frac{1}{\sqrt{e^{2x}-1}} \frac{d}{dx}\left(e^x\right) = \frac{e^x}{\sqrt{e^{2x}-1}}$$

Solved Problems

Solved Problem 4.1 Find the first derivative of:

$$y = \tan x^2$$

Solution.

$$y' = \sec^2 x^2 \frac{d}{dx}\left(x^2\right) = 2x \sec^2 x^2$$

Solved Problem 4.2 Find the first derivative of:

$$y = \tan^2 x = (\tan x)^2$$
Solution.

$$y' = 2 \tan x \frac{d}{dx}\left(\tan x\right) = 2 \tan x \sec^2 x$$

Solved Problem 4.3 Find the derivative of:

$y = x - \sin x \cos x$

Solution.

$$\frac{dy}{dx} = 1 - \frac{d}{dx}(\sin x \cos x)$$

$$\frac{d}{dx}(\sin x \cos x) = \cos x \cos x + (-\sin x)\sin x$$

$$= \cos^2 x - \sin^2 x$$

Therefore, by substituting into the previous equation, the derivative can be found:

$$\frac{dy}{dx} = 1 - (\cos^2 x - \sin^2 x) = 1 - \cos^2 x + \sin^2 x$$

Solved Problem 4.4 Find the derivative of:

$$\frac{d}{dx}\left(\frac{\csc x}{\sqrt{x}}\right)$$

Solution.

$$\frac{dy}{dx} = 1 - \frac{d}{dx}(\sin x \cos x)$$

$$\frac{d}{dx}(\sin x \cos x) = \cos x \cos x + (-\sin x)\sin x$$

$$= \cos^2 x - \sin^2 x$$

Simplifying leads to:

$$\frac{d}{dx}\left(\frac{\csc x}{\sqrt{x}}\right) = \frac{-\csc x \cot x}{\sqrt{x}} - \frac{\csc x}{2x^{3/2}}$$

Solved Problem 4.5 Find the first derivative of:

$$y = x^2 3^x$$

Solution.

$$y' = x^2 \frac{d}{dx}(3^x) + 3^x \frac{d}{dx}(x^2) = x^2 3^x \ln 3 + 3^x 2x = x3^x(x \ln 3 + 2)$$

Solved Problem 4.6 Find the derivative of:

$$f(x) = \sin^2 3x$$

Solution. The derivative of the function is the composite of the functions below:

$$y = u^2, \ u = \sin v, \ v = 3x$$

Therefore,

$$\frac{d}{dx}(\sin^2 3x) = \frac{d}{dx}(u^2)\frac{d}{dx}(\sin v)\frac{d}{dx}(3x)$$
$$= (2u)(\cos v)(3)$$
$$= 6u \cos v$$

Making the appropriate substitutions yields:

$$\frac{d}{dx}(\sin^2 3x) = 6 \sin v \cos v$$
$$= 6 \sin 3x \cos 3x$$

Solved Problem 4.7 Find the derivative of:

$$y = \sin^{-1}\left(\frac{5x}{6}\right)$$

Solution.

The function above is a composite of two functions and thus the derivative is determined by using the chain rule, where

$$y = \sin^{-1} u \qquad \text{where} \quad u = \frac{5x}{6}$$

$$\frac{dy}{du} = \frac{1}{\sqrt{1-u^2}} \qquad \text{and} \qquad \frac{du}{dx} = \frac{5}{6}$$

Therefore,

$$\frac{dy}{dx} = \frac{dy}{du}\frac{du}{dx}$$

$$= \left(\frac{1}{\sqrt{1-u^2}}\right)\left(\frac{5}{6}\right) = \left(\frac{1}{\sqrt{1-\left(\frac{5}{6}x\right)^2}}\right)\left(\frac{5}{6}\right) = \left(\frac{1}{\sqrt{1-\left(\frac{25}{36}x^2\right)}}\right)\left(\frac{5}{6}\right)$$

$$= \left(\frac{\sqrt{36}}{\sqrt{36-25x^2}}\right)\left(\frac{5}{6}\right)$$

$$= \left(\frac{5}{\sqrt{36-25x^2}}\right)$$

Solved Problem 4.8 Find the derivative of:

$$y = \sec^{-1} 6x$$

Solution. This problem requires the chain rule given

$$y = \sec^{-1} u \quad \text{where} \quad u = 6x$$

Thus,

$$\frac{dy}{dx} = \frac{1}{|u|\sqrt{u^2-1}} \cdot 6 = \frac{1}{|6x|\sqrt{36\,x^2-1}} \cdot 6$$

$$= \frac{1}{|6||x|\sqrt{36\,x^2-1}} \cdot 6 = \frac{1}{|x|\sqrt{36\,x^2-1}}$$

Chapter 5
THE LAW OF THE MEAN, INDETERMINATE FORMS, DIFFERENTIALS, AND CURVE SKETCHING

IN THIS CHAPTER:

- ✔ Rolle's Theorem
- ✔ The Law of the Mean
- ✔ Indeterminate Forms
- ✔ Differentials
- ✔ Curve Sketching
- ✔ Solved Problems

Rolle's Theorem

If f(x) is continuous on the interval $a \leq x \leq b$, if f(a) = f(b) = 0, and if f '(x) exists everywhere on the interval except possibly at the endpoints, then f '(x) = 0 for at least one value of x, say $x = x_0$, between a and b.

Geometrically, this means that if a continuous curve intersects the x axis at x = a and x = b, and has a tangent at every point between a and b, then

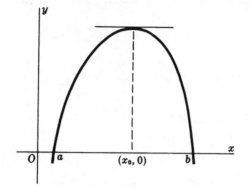

Figure 5-1

Corollary

If f(x) satisfies the conditions of Rolle's theorem, except that f(a) = f(b) ≠ 0, then f′(x) = 0 for at least one value of x, say x = x_o, between a and b. (See Figure 5-2.) This is to say, if the line containing the endpoints is horizontal (and therefore has zero slope), then the slope of the tangent is also zero for some point in between.

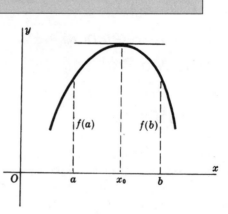

Figure 5-2

there is at least one point $x = x_0$ between a and b where the tangent is parallel to the x axis. (See Figure 5-1.)

Example 5.1 Find the value of x_0 prescribed in Rolle's theorem for $f(x) = x^3 - 12x$ on the interval $0 \le x \le 2\sqrt{3}$.

$f'(x) = 3x^2 - 12 = 0$ when $x = \pm 2$; then $x_0 = 2$ in the prescribed value.

The Law of the Mean

If $f(x)$ is continuous on the interval $a \le x \le b$, and if $f'(x)$ exists everywhere on the interval except possibly at the endpoints, then there is at least one value $x = x_0$, between a and b, such that

$$\frac{f(b)-f(a)}{b-a} = f'(x_0)$$

Also known as the mean-value theorem, this means, geometrically, that if P_1 and P_2 are two points of a continuous curve that has a tangent at each intermediate point between P_1 and P_2, then there exists at least one point of the curve between P_1 and P_2 at which the slope of the curve is equal to the slope of the line between the endpoints, P_1 and P_2. (See Figure 5-3.)

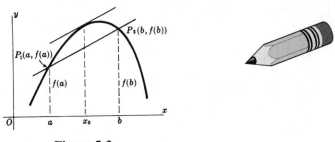

Figure 5-3

The law of the mean may be put in several useful forms. The first is obtained by multiplication by b - a:

$f(b) = f(a) + (b - a)f'(x_0)$ for some x_0 between a and b   (5.1)

A simple change of letter yields an expression for any arbitrary value of x:

$$f(x) = f(a) + (x - a)f'(x_o) \quad \text{for some } x_o \text{ between a and x} \tag{5.2}$$

Example 5.2 Use the law of the mean to approximate $\sqrt[6]{65}$.

Let $f(x) = \sqrt[6]{x}$, a = 64, and b = 65, and apply Eq. (5.1), obtaining

$$f(65) = f(64) + \frac{65 - 64}{6x_o^{5/6}}, \quad 64 < x_o < 65$$

Since x_o is not known, take $x_o = 64$; then approximately,

$$\sqrt[6]{65} = \sqrt[6]{64} + 1/\left(6\sqrt[6]{64^5}\right) = 2 + 1/192 = 2.00521$$

Example 5.3 A circular hole with a diameter of 4 in and a depth of 1 ft in a metal block is rebored to increase the diameter to 4.12 in. Estimate the amount of metal removed.

The volume of a circular hole of radius x in and depth 12 in is given by $V = f(x) = 12 \pi x^2$. We are to estimate f(2.06) - f(2). By the law of the mean,

$$f(2.06) - f(2) = 0.06 \, f'(x_o) = 0.06(24 \, \pi \, x_o) \qquad 2 < x_o < 2.06$$

Take $x_o = 2$; then, approximately,

$$f(2.06) - f(2) = 0.06(24 \, \pi)(2) = 2.88 \, \pi \, in^3.$$

Generalized Law of the Mean

If f(x) and g(x) are continuous on the interval $a \le x \le b$, and if f'(x) and g'(x) exist and $g'(x) \ne 0$ everywhere on the interval except possibly at the endpoints, then there exists at least one value of x, say $x = x_o$, between a and b such that

$$\frac{f(b) - f(a)}{g(b) - g(a)} = \frac{f'(x_o)}{g'(x_o)}$$

For the case g(x) = x, this becomes the law of the mean.

Extended Law of the Mean

If f(x) and its first n - 1 derivatives are continuous on the interval $a \leq x \leq b$, and if $f^{(n)}(x)$ exists everywhere on the interval except possibly at the endpoints, then there exists at least one value of x, say $x = x_0$, between a and b such that

$$f(b)=f(a)+\frac{f'(a)}{1!}(b-a)+\frac{f''(a)}{2!}(b-a)^2+\cdots$$

$$+\frac{f^{(n-1)}(a)}{(n-1)!}(b-a)^{n-1}+\frac{f^{(n)}(x_0)}{n!}(b-a)^n \tag{5.3}$$

When b is replaced with the variable x, Eq. (5.3) becomes

$$f(x)=f(a)+\frac{f'(a)}{1!}(x-a)+\frac{f''(a)}{2!}(x-a)^2+\cdots$$

$$+\frac{f^{(n-1)}(a)}{(n-1)!}(x-a)^{n-1}+\frac{f^{(n)}(x_0)}{n!}(x-a)^n \tag{5.4}$$

for some x_0 between a and x.

When a is replaced with 0, Eq. (5.4) becomes

$$f(x)=f(0)+\frac{f'(0)}{1!}x+\frac{f''(0)}{2!}x^2+\cdots$$

$$+\frac{f^{(n-1)}(0)}{(n-1)!}x^{n-1}+\frac{f^{(n)}(x_0)}{(n)!}x^n \tag{5.5}$$

for some x_0 between 0 and x.

Indeterminate Forms

The derivative of a differentiable function f(x) is defined as:

$$\lim_{\Delta x \to 0} \frac{f(x+\Delta x)-f(x)}{(x+\Delta x)-x} \tag{5.6}$$

Since the limit of both the numerator and the denominator of the fraction is zero, Eq. (5.6) is an example of a limit which is called **indeterminate** of the type 0/0. Similarly, it is customary to call a limit such as

$$\lim_{x\to\infty} \frac{3x-2}{9x+7}$$

indeterminate of the type ∞/∞. These symbols, 0/0, ∞/∞, and others $(0\cdot\infty, \infty - \infty, 0^0, \infty^0, \text{ and } 1^\infty)$ to be introduced later must not be taken literally; they have no numerical meaning and are merely convenient labels for distinguishing types of behavior at certain limits.

Indeterminate Type 0/0; L'Hospital's Rule

If a is a number, if f(x) and g(x) are differentiable and $g(x) \neq 0$ for all x on some interval $0 < |x - a| < \delta$, and if $\lim_{x\to a} f(x) = 0$ and $\lim_{x\to a} g(x) = 0$,

then, when

$$\lim_{x\to a} \frac{f'(x)}{g'(x)} \quad \text{exists or is infinite, we can write}$$

$$\lim_{x\to a} \frac{f(x)}{g(x)} = \lim_{x\to a} \frac{f'(x)}{g'(x)} \qquad \textbf{(L'Hospital's rule)}$$

Example 5.4

$$\lim_{x\to 3} \frac{x^4-81}{x-3} = 108 \quad \text{is indeterminate of type 0/0. Because}$$

$$\lim_{x\to 3} \frac{\dfrac{d}{dx}(x^4-81)}{\dfrac{d}{dx}(x-3)} = \lim_{x\to 3} 4x^3 = 108$$

we have $\lim\limits_{x\to 3}\ \dfrac{x^4-81}{x-3}$

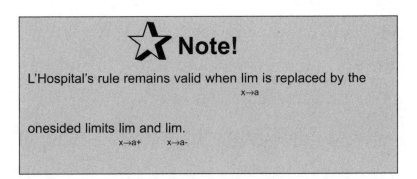

⭐ Note!

L'Hospital's rule remains valid when $\lim\limits_{x\to a}$ is replaced by the

onesided limits $\lim\limits_{x\to a^+}$ and $\lim\limits_{x\to a^-}$.

Indeterminate Type ∞/∞

The conclusion of l'Hospital's rule is unchanged if one or both of the following changes are made in the hypotheses:

1. "$\lim\limits_{x\to a}$ f(x) = 0 and $\lim\limits_{x\to a}$ g(x) = 0" is replaced by

   "$\lim\limits_{x\to a}$ f(x) = ∞ and $\lim\limits_{x\to a}$ g(x) = ∞."

2. "a is a number" is replaced by "a = + ∞, - ∞, or ∞" and
   "0 < |x - a| < δ" is replaced by "|x| > M."

Example 5.5

$\lim\limits_{x\to+\infty}\ \dfrac{x^2}{e^x}$ is indeterminate of type ∞/∞. Applying l'Hospital's rule

twice gives us

$\lim\limits_{x\to+\infty}\ \dfrac{x^2}{e^x} = \lim\limits_{x\to+\infty}\ \dfrac{2x}{e^x} = \lim\limits_{x\to+\infty}\ \dfrac{2}{e^x} = 0$

Indeterminate Types $0 \cdot \infty$ and $\infty - \infty$

These may be handled by first transforming to one of the types 0/0 or ∞/∞. For example,

$\lim\limits_{x \to +\infty} x^2 e^{-x}$ is of type $0 \cdot \infty$ but

$\lim\limits_{x \to +\infty} \dfrac{x^2}{e^x}$ is of type ∞/∞.

$\lim\limits_{x \to +0} \left(\csc x - \dfrac{1}{x} \right)$ is of type $\infty - \infty$ but

$\lim\limits_{x \to +0} \left(\dfrac{x - \sin x}{x \sin x} \right)$ is of type 0/0.

Example 5.6 Evaluate $\lim\limits_{x \to 0+} (x^2 \ln x)$.

As $x \to 0^+$, $x^2 \to 0$ and $\ln x \ - \infty$. Then, $\dfrac{\ln x}{1/x^2}$ has an indeterminate limit of type ∞/∞.

$$\lim\limits_{x \to 0+} (x^2 \ln x) = \lim\limits_{x \to 0+} \dfrac{\ln x}{1/x^2} = \lim\limits_{x \to 0+} \dfrac{1/x}{-2/x^3} = \lim\limits_{x \to 0+} \left(-\dfrac{1}{2} x^2 \right) = 0$$

Indeterminate Types 0^0, ∞^0, and 1^∞

If $\lim y$ is one of these types, then $\lim (\ln y)$ is of the type $0 \cdot \infty$.

Example 5.7 Evaluate $\lim\limits_{x \to 0} \left(\sec^3 2x \right)^{\cot^2 3x}$.

This is of the type 1^∞. Let $y = \left(\sec^3 2x \right)^{\cot^2 3x}$.

Then, $\ln y = \cot^3 3x \ln \sec^3 2x = \dfrac{3 \ln \sec 2x}{\tan^2 3x}$

and $\lim\limits_{x \to 0} \ln y$ is of the type 0/0. L'Hospital's rule gives

$$\lim_{x \to 0} \frac{3 \ln \sec 2x}{\tan^2 3x} = \lim_{x \to 0} \frac{6 \tan 2x}{6 \tan 3x \, \sec^2 3x} = \lim_{x \to 0} \frac{\tan 2x}{\tan 3x}$$

since $\lim\limits_{x \to 0} \sec^2 3x = 1$, and the last limit above is of the type 0/0.

L'Hospital's rule now gives

$$\lim_{x \to 0} \frac{\tan 2x}{\tan 3x} = \lim_{x \to 0} \frac{2 \sec^2 2x}{3 \sec^2 3x} = \frac{2}{3}$$

Since $\lim\limits_{x \to 0} \ln y = \dfrac{2}{3}$, then $\lim\limits_{x \to 0} y = \lim\limits_{x \to 0} \left(\sec^3 2x \right)^{\cot^2 3x} = e^{2/3}$

Differentials

For the function $y = f(x)$, we define the following:

1. dx, called the **differential of x**, given by the relation $dx = \Delta x$
2. dy, called the **differential of y**, given by the relation $dy = f'(x)dx$

The differential of the independent variable is, by definition, equal to the increment of the variable. But the differential of the dependent variable is *not* equal to the increment of that variable. (See Figure 5-4.)

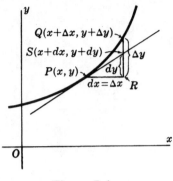

Figure 5-4

The increment Δy measures the vertical displacement from an initial point x_0 if followed along the curve $y = f(x)$, whereas the differential dy measures the vertical displacement from x_0 if followed along the tangent to the curve at x_0.

Example 5.8 When $y = x^2$, $dy = 2x\,dx$ while $\Delta y = (x + \Delta x)^2 - x^2 = 2x\,\Delta x + (\Delta x)^2 = 2x\,dx + (dx)^2$. A geometric interpretation is given in Figure 5-5, where it can be seen that Δy and dy differ by the small square of area $(dx)^2$.

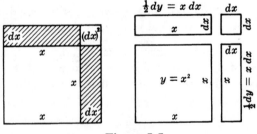

Figure 5-5

The differential dy may be found by using the definition $dy = f'(x)dx$ or by means of rules obtained readily from the rules for finding derivatives. Some of these are:

$$d(c)=0 \qquad\qquad d(cu)=c\ du \qquad d(uv)=u\ dv+v\ du$$

$$d\left(\frac{u}{v}\right)=\frac{v\ du - u\ dv}{v^2} \qquad d(\sin u)=\cos u\ du \qquad d(\ln u)=\frac{du}{u}$$

Example 5.9 Find dy for each of the following:

(a) $y = x^3 + 4x^2 - 5x + 6$

$dy = d(x^3) + d(4x^2) - d(5x) + d(6) = (3x^2 + 8x - 5)dx$

(b) $y = (2x^3 + 5)^{3/2}$

$$dy=\frac{3}{2}\left(2x^3+5\right)^{1/2}d\left(2x^3+5\right)=\frac{3}{2}\left(2x^3+5\right)^{1/2}\left(6x^2\ dx\right)=9x^2\left(2x^3+5\right)^{1/2}dx$$

Approximations by Differentials

If dx = Δx is relatively small when compared with x, dy is a fairly good approximation of Δy, that is, $\lim\limits_{\Delta x \to 0} \Delta y = dy$.

Example 5.10 Take $y = x^2 + x + 1$, and let x change from x = 2 to x = 2.01. The actual change in y is $\Delta y = [(2.01)^2 + 2.01 + 1] - (2^2 + 2 + 1) = 0.0501$. The approximate change in y, obtained by taking x = 2 and dx = 0.01, is dy = f'(x) dx = (2x + 1) dx = [2(2) + 1](0.01) = 0.05.

Approximations of Roots of Equations

Let $x = x_1$ be a fairly close approximation of a root r of the equation $y = f(x) = 0$, and let $f(x_1) = y_1 \neq 0$. Then y_1 differs from 0 by a small amount. Now if x_1 were changed to r, the corresponding change in $f(x_1)$ would be $\Delta y_1 = -y_1$. An approximation of this change in x_1 is given by $f'(x_1)dx_1 = -y_1$ or

$$dx_1 = -\frac{y_1}{f'(x_1)}$$

Thus, a second and better approximation of the root r is

$$x_2 = x_1 + dx_1 = x_1 - \frac{y_1}{f'(x_1)} = x_1 - \frac{f(x_1)}{f'(x_1)}$$

A third approximation is

$$x_3 = x_2 + dx_2 = x_2 - \frac{f(x_2)}{f'(x_2)}$$

and so on. (See Figure 5-6)

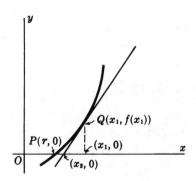

Figure 5-6

When x_1 is not a sufficiently close approximation of a root, it will be found that x_2 differs materially from x_1. While at times the process of finding these approximations is self-correcting, it is often simpler to make a new first approximation.

Example 5.11 Approximate the roots of $2 \cos x - x^2 = 0$.

The curves $y = 2 \cos x$ and $y = x^2$ intersect in two points whose abscissas are approximately 1 and - 1. (Note that if r is one root, then -r is the other.)

Using $x_1 = 1$ yields

$$x_2 = 1 - \frac{2 \cos 1 - 1}{-2 \sin 1 - 2} = 1 + \frac{2(0.5403) - 1}{2(0.8415) + 2} = 1 + 0.02 = 1.02$$

Then,

$$x^3 = 1.02 - \frac{2 \cos(1.02) - (1.02)^2}{-2 \sin(1.02) - 2(1.02)} = 1.02 + \frac{0.0064}{3.7442}$$

$$= 1.02 + 0.0017 = 1.0217$$

Thus, to four decimal places, the roots are 1.0217 and -1.0217.

Curve Sketching

Symmetry

A curve is **symmetric** with respect to:

1. The x axis, if its equation is unchanged when y is replaced by -y, that is, $f(x) = y$ and $f(x) = -y$ both hold.

2. The y axis, if its equation is unchanged when x is replaced by -x, that is $f(-x) = f(x)$.

3. The origin, if its equation is unchanged when x is replaced by -x and y by -y, simultaneously, that is, $f(-x) = -f(x)$.

4. The line $y = x$, if its equation is unchanged when x and y are interchanged, that is, $y = f(x)$ implies $x = f(y)$.

Intercepts

An **intercept** is a point on either axis of a coordinate system where the curve of a function passes through or intercepts. The x intercepts are obtained by setting $y = 0$ in the equation for the curve and solving for x (when possible). The y intercepts are obtained by setting $x = 0$ and solving for y.

Extent

The **horizontal extent** of a curve is given by the values of x, for which the curve exists. The **vertical extent** is given by the range of y. A point (x_o, y_o) is called an **isolated point** of a curve if its coordinates satisfy the equation of the curve while those of no other nearby point do.

Asymptotes

An **asymptote** of a curve is a line that comes arbitrarily close to the curve as the abscissa or ordinate of the curve approaches infinity. Specifically, given a curve $y = f(x)$, the vertical asymptotes $x = a$ can be described by the infinite limits: $\lim_{x \to a} f(x) = \pm \infty$. Likewise, the horizontal asymptotes $y = b$ can be defined as the limits at infinity: $\lim_{x \to \pm\infty} f(x) = b$.

The maximum and minimum points, points of inflection, and concavity of a curve are discussed in Chapter 4.

Example 5.12 Discuss and sketch the curve $y^2(1 + x) = x^2(1 - x)$.

The curve is drawn in Figure 5-7.

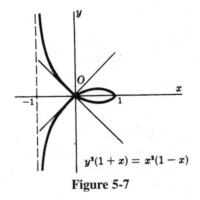

$$y^2(1 + x) = x^2(1 - x)$$

Figure 5-7

We may write the equation of the curve as

$$y^2 = \frac{x^2(1-x)}{1+x}$$

Symmetry: The curve is symmetric with respect to the x axis.

Intercepts: The x intercepts are x = 0 and x = 1. The y intercept is y = 0.

Extent: For x = 1, y = 0. For x = -1, there is no point on the curve. For other values of x, y^2 must be positive so 1 + x and 1 - x must have the same sign; hence, for points on the curve, x is restricted to -1 < x < 1. Thus, -1 < x ≤ 1.

Asymptotes: $$y^2 = \frac{x^2(1-x)}{1+x}$$

Hence, y →∞ as x →-1. Thus, x = -1 is a vertical asymptote.

Maximum and minimum points, etc.: The curve consists of two branches

$$y = \frac{x\sqrt{1-x}}{\sqrt{1+x}} \quad \text{and} \quad y = -\frac{x\sqrt{1-x}}{\sqrt{1+x}}$$

For the first of these,

$$\frac{dy}{dx} = \frac{1-x-x^2}{(1+x)^{3/2}(1-x)^{1/2}} \quad \text{and} \quad \frac{d^2y}{dx^2} = \frac{x-2}{(1+x)^{5/2}(1-x)^{3/2}}$$

The critical values are x = 1 and $(-1 + \sqrt{5})/2$. The point

$$\left(\frac{-1+\sqrt{5}}{2}, \frac{(-1+\sqrt{5})\sqrt{\sqrt{5}-2}}{2} \right)$$

is a maximum point. There is no point of inflection. The branch is concave downward. By symmetry, there is a minimum point at

$$\left(\frac{-1+\sqrt{5}}{2}, \frac{\left(-1+\sqrt{5}\right)\sqrt{\sqrt{5}-2}}{2}\right)$$

and the second branch is concave upward.

The curve passes through the origin twice. The tangent lines at the origin are the lines y = x and y = -x.

Example 5.13 Discuss and sketch the curve $y=\dfrac{\ln x}{x}$.
The curve is drawn in Figure 5-8.

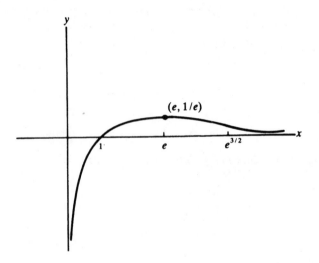

Figure 5-8

Symmetry: There is no symmetry.

Intercepts: The only x intercept is x = 1.

Extent: The curve is defined for x > 0.

Asymptotes: The y axis is a vertical asymptote,

$\dfrac{\ln x}{x} \to -\infty$ since as x—>0⁺. By l'Hospital's rule,

$\dfrac{\ln x}{x} \to 0$

as x —> +∞. Hence, the positive x axis is a horizontal asymptote, that is, the line y = 0.

Maximum and minimum points, etc.: We have

$$\dfrac{dy}{dx} = \dfrac{1-\ln x}{x^2} \quad \text{and} \quad \dfrac{d^2y}{dx^2} = \dfrac{2\ln x - 3}{x^3}$$

Hence, the critical point is (e, 1/e). At that point,

$$\dfrac{d^2y}{dx^2} = -\dfrac{1}{e^3} < 0$$

which gives us a relative maximum. There is a point of inflection for 2 ln x = 3, that is, at (e^{3/2}, 3/2e^{3/2}). The curve is concave downward for 0 < x < e^{3/2} and concave upward for x > e^{3/2}.

Solved Problems

Solved Problem 5.1 Verify the mean value theorem for the function:

y = 4x³ - x + 5

on the interval [1, 4].

Solution. We first calculate the function value at the endpoints of the interval a = 1 and b = 4.

f(a) = f(1) = 4(1)³ - 1 + 5 = 4 - 1 + 5 = 8

f(b) = f(4) = 4(4)³ - 4 + 5 = 256 - 4 + 5 = 257

According to the mean value theorem, there exists one number, c, in the open interval (a, b) such that

$$f'(c) = \frac{f(b) - f(a)}{b - a} = \frac{257 - 8}{4 - 1} = \frac{249}{3} = 83$$

To find c explicitly, we determine the derivative of the original function defined at c:

$$f'(x) = 12x^2 - 1 = 83$$

This can be simplified:

$$12x^2 = 84$$

$$x^2 = 7$$

$$x = \pm\sqrt{7}$$

Since only $+\sqrt{7}$ is in (1, 4), it is the number that serves as the c whose existence is guaranteed by the mean value theorem.

Solved Problem 5.2 In graphing a function, what holds true if (a) the ordinate f(x), (b) the slope f'(x), and (c) the second derivative f"(x) are positive?

Solution. (a) When the ordinate f(x) is positive, the graph is above the x-axis.

(b) When the slope f"(x) is positive, the graph slopes upward.

(c) When the second derivative f"(x) is positive, the graph is concave upward.

Solved Problem 5.3 Repeat Solved Problem 5.2, assuming all entities described are negative?

Solution. (a) When the ordinate f(x) is negative, the graph is below the x-axis.

(b) When the slope $f''(x)$ is negative, the graph slopes downward.
(c) When the second derivative $f''(x)$ is positive, the graph is concave downward.

Solved Problem 5.4 Repeat Solved Problem 5.2, assuming all entities described change sign?

Solution. (a) When the ordinate $f(x)$ changes sign, the graph crosses the x-axis.

(b) When the slope $f''(x)$ changes sign, the graph has a horizontal tangent and a relative maximum or minimum.

(c) When the second derivative $f''(x)$ changes sign, the graph has an inflection point.

Chapter 6
FUNDAMENTAL INTEGRATION TECHNIQUES AND APPLICATIONS

IN THIS CHAPTER:

✔ *Fundamental Integration Formulas*
✔ *Integration by Parts*
✔ *Trigonometric Integrals*
✔ *Trigonometric Substitutions*
✔ *Integration by Partial Fractions*
✔ *Miscellaneous Substitutions*
✔ *Other Substitutions*
✔ *Integration of Hyperbolic Functions*
✔ *Applications of Indefinite Integrals*
✔ *Solved Problems*

If F(x) is a function whose derivative F'(x) = f(x) on a certain interval of the x axis, then F(x) is called an **antiderivative** or **indefinite integral** of f(x). The indefinite integral of a given function is not unique; for example, x^2, $x^2 + 5$, and $x^2 - 4$ are all indefinite integrals of f(x) = 2x, since

$$\frac{d}{dx}\left(x^2\right)=\frac{d}{dx}\left(x^2+5\right)=\frac{d}{dx}\left(x^2-4\right)=2x$$

All indefinite integrals of $f(x) = 2x$ are then described by the general form of the antiderivative $F(x) = x^2 + C$, where C, called the constant of integration, is an arbitrary constant.

The symbol $\int f(x)dx$ is used to indicate the indefinite integral of $f(x)$ where the function $f(x)$ is called the **integrand**. Thus we write

$$\int 2x\ dx = x^2+C$$

where dx denotes the antiderivative being taken with respect to x.

Fundamental Integration Formulas

A number of the formulas below follow immediately from the standard differentiation formulas of earlier sections, while others may be checked by differentiation. Formula 25 displayed below, for example, may be checked by showing that

$$\frac{d}{dx}\left(\frac{1}{2}x\sqrt{a^2-x^2}+\frac{1}{2}a^2\arcsin\frac{x}{a}+C\right)=\sqrt{a^2-x^2}$$

$$\int 2x\ dx=x^2+C$$

Absolute value signs appear in several of the formulas. For example, for Formula 5 displayed below, we write

$$\int\frac{dx}{x}=\ln\ |x|+C$$

instead of

$$\int\frac{dx}{x}=\ln\ x+C\ \text{ for } x > 0 \ \text{ and } \int\frac{dx}{x}=\ln\ (-x)+C\ \text{ for } x < 0$$

1. $\int \dfrac{d}{dx}[f(x)]dx = f(x) + C$ · (Fundamental Theorem of Calculus)

2. $\int [f(x) + g(x)]dx = \int f(x)dx + \int g(x)dx$

3. $\int af(x)dx = a \int f(x)dx$, $a = $ any constant

4. $\int x^m dx = \dfrac{x^{m+1}}{m+1} + C$, $m \neq -1$

5. $\int \dfrac{dx}{x} = \ln |x| + C$

6. $\int a^x dx = \dfrac{a^x}{\ln a} + C$, $a > 0$, $a \neq 1$

7. $\int e^x dx = e^x + C$

8. $\int \sin x \, dx = -\cos x + C$

9. $\int \cos x \, dx = \sin x + C$

10. $\int \tan x \, dx = \ln |\sec x| + C$

11. $\int \cot x \, dx = \ln |\sin x| + C$

12. $\int \sec x \, dx = \ln |\sec x + \tan x| + C$

13. $\int \csc x \, dx = \ln |\csc x - \cot x| + C$

14. $\int \sec^2 x \, dx = \tan x + C$

15. $\int \csc^2 x \, dx = -\cot x + C$

16. $\int \sec x \tan x \, dx = \sec x + C$

17. $\int \csc x \cot x \, dx = -\csc x + C$

18. $\int \dfrac{dx}{\sqrt{a^2 - x^2}} = \arcsin \dfrac{x}{a} + C$

19. $\int \dfrac{dx}{a^2 + x^2} = \dfrac{1}{a} \arctan \dfrac{x}{a} + C$

20. $\int \dfrac{dx}{x \sqrt{x^2 - a^2}} = \dfrac{1}{a} \operatorname{arcsec} \dfrac{x}{a} + C$

21. $\int \dfrac{dx}{x^2 - a^2} = \dfrac{1}{2a} \ln \left| \dfrac{x - a}{x + a} \right| + C$

22. $\int \dfrac{dx}{a^2 - x^2} = \dfrac{1}{2a} \ln \left| \dfrac{a + x}{a - x} \right| + C$

23. $\int \dfrac{dx}{\sqrt{x^2 + a^2}} = \ln \left(x + \sqrt{x^2 + a^2} \right) + C$

24. $\int \dfrac{dx}{\sqrt{x^2 - a^2}} = \ln \left| x + \sqrt{x^2 - a^2} \right| + C$

25. $\int \sqrt{a^2 - x^2} \, dx = \dfrac{1}{2} x \sqrt{a^2 - x^2} + \dfrac{1}{2} a^2 \arcsin \dfrac{x}{a} + C$

26. $\int \sqrt{x^2 + a^2} \, dx = \dfrac{1}{2} x \sqrt{x^2 + a^2} + \dfrac{1}{2} a^2 \ln \left(x + \sqrt{x^2 + a^2} \right) + C$

The Method of Substitution

To evaluate an antiderivative $\int f(g(x)) dx \cdot g'(x) dx$, it is often useful to replace g(x) with a new variable u by means of a **substitution** u = g(x), du = g'(x) dx. The equation

$$\int f(u) \, du = \int f(g(x)) g'(x) \, dx \tag{6.1}$$

is valid. After finding the right side of Eq. (6.1), we replace u with g(x); that is, we obtain the result in the original terms of x. To verify Eq. (6.1), observe that, if

$F(x) = \int f(x)dx$, then

$$\frac{d}{du}F(x) = \frac{d}{dx}F(x)\frac{dx}{du} = f(x)g'(u) = f(g(u))g'(u)$$

Hence,

$$F(x) = \int f(g(u))g'(u)du$$

which is Eq. (6.1). This method, called "u-substitution" applies to integrands which are a product of the form Eq. (6.1). (Such a product results from the chain rule applied to the original composite function F(g(x)).)

Example 6.1 Evaluate $\int (x+3)^{11}dx$.

To evaluate the integral, replace x + 3 with u; that is, let u = x + 3. Then dx = du, and we obtain

$$\int (x+3)^{11}dx = \int u^{11}du = \frac{1}{12}u^{12} + C = \frac{1}{12}(x+3)^{12} + C$$

Quick Integration by Inspection

Two simple formulas enable us to find antiderivatives almost immediately. The first is

$$\int g'(x)[g(x)]^r dx = \frac{1}{r+1}[g(x)]^{r+1} + C, \; r \neq -1 \tag{6.2}$$

This formula is justified by noting that

$$\frac{d}{dx}\left\{ \frac{1}{r+1}[g(x)]^{r+1} \right\} = g'(x)[g(x)]^r$$

Examples 6.2 Evaluate the following integrals

(a) $\int \dfrac{(\ln x)^2}{x} dx$ and (b) $\int x \sqrt{x^2+3}\ dx$

(a) $\int \dfrac{(\ln x)^2}{x} dx = \int \dfrac{1}{x}(\ln x)^2 dx = \dfrac{1}{3}(\ln x)^3 + C$ (verify by differentiation)

(b) $\int x\sqrt{x^2+3}\ dx = \dfrac{1}{2}\int (2x)(x^2+3)^{1/2} dx = \dfrac{1}{2}\left[\dfrac{1}{3/2}(x^2+3)^{3/2}\right]+C$

$$= \dfrac{1}{3}\left[\sqrt{x^2+3}\ \right]^3 + C$$

The second quick integration formula is

$$\int \dfrac{g'(x)}{g(x)} dx = \ln |g(x)| + C \tag{6.3}$$

This formula is justified by noting that

$$\dfrac{d}{dx}\left(\ln |g(x)|\right) = \dfrac{g'(x)}{g(x)}$$

Examples 6.3 Evaluate the following integrals

(a) $\int \cot x\ dx$ and (b) $\int \dfrac{x^2}{x^3-5} dx$

(a) $\int \cot x\ dx = \int \dfrac{\cos x}{\sin x} dx = \ln |\sin x| + C$

(b) $\int \dfrac{x^2}{x^3-5} dx = \dfrac{1}{3}\int \dfrac{3x^2}{x^3-5} dx = \dfrac{1}{3}\ln |x^3-5| + C$

Integration by Parts

When u and v are differentiable functions of x, then

$$d(uv) = u\ dv + v\ du \text{ or } u\ dv = d(uv) - v\ du \tag{6.4}$$

and, integrating both sides, we get

$$\int u\ dv = uv - \int v\ du \tag{6.5}$$

When used in a required integration, the given integral must be separated into two parts, one part being u and the other part, together with dx, being dv. For this reason, integration by use of Eq. (6.5) is called **integration by parts**. Two general rules can be stated:

1. The part selected as dv must be readily integrable.

2. $\int v \, du$ must not be more complex than $\int v \, du$.

Example 6.4 Find $\int x^3 e^{x^2} dx$.

Take u = x^2 and dv = $e^{x^2} x \, dx$; then du = 2x dx and v = $\dfrac{1}{2} e^{x^2}$ (by u-substitution). Now, by Eq. (6.5),

$$\int x^3 e^{x^2} dx = \frac{1}{2} x^2 e^{x^2} - \int x e^{x^2} dx = \frac{1}{2} x^2 e^{x^2} - \frac{1}{2} e^{x^2} + C$$

(where again, we applied u-substitution to find the integral $\int x e^{x^2} dx$).

Example 6.5 Find $\int \ln(x^2 + 2) dx$.

Take u = ln $(x^2 + 2)$ and dv = dx; then $du = \dfrac{2x \, dx}{x^2 + 2}$ and v = x. By Eq. (6.5),

$$\int \ln(x^2 + 2) dx = x \ln(x^2 + 2) - \int \frac{2x^2}{x^2 + 2} dx$$

$$= x \ln(x^2 + 2) - \int \left(2 - \frac{4}{x^2 + 2} \right) dx$$

$$= x \ln(x^2 + 2) - 2x + 2\sqrt{2} \arctan \frac{x}{\sqrt{2}} + C$$

Reduction Formulas

The labor involved in successive applications of integration by parts to evaluate an integral may be materially reduced by the use of **reduction formulas**.

> ### ★ Note!
>
> In general, a reduction formula yields a new integral of the same form as the original but with an exponent increased or reduced. A reduction formula succeeds if ultimately it produces an integral that can be evaluated. Among the reduction formulas are:

$$\int \frac{dx}{\left(a^2 \pm x^2\right)^m} = \frac{1}{a^2}\left[\frac{x}{(2m-2)\left(a^2 \pm x^2\right)^{m-1}} + \frac{2m-3}{2m-2}\int \frac{dx}{\left(a^2 \pm x^2\right)^{m-1}}\right],$$

$$m \neq 1 \qquad (6.7)$$

$$\int \left(a^2 \pm x^2\right)^m dx = \frac{x\left(a^2 \pm x^2\right)^m}{2m+1} + \frac{2ma^2}{2m+1}\int \left(a^2 \pm x^2\right)^{m-1} dx, \; m \neq -\frac{1}{2} \qquad (6.8)$$

$$\int \frac{dx}{\left(x^2 - a^2\right)^m} = -\frac{1}{a^2}\left[\frac{x}{(2m-2)\left(x^2 - a^2\right)^{m-1}} + \frac{2m-3}{2m-2}\int \frac{dx}{\left(x^2 - a^2\right)^{m-1}}\right],$$

$$m \neq 1 \qquad (6.9)$$

$$\int \left(x^2 - a^2\right)^m dx = \frac{x\left(x^2 - a^2\right)^m}{2m+1} - \frac{2ma^2}{2m+1}\int \left(x^2 - a^2\right)^{m-1} dx, \;\; m \neq -\frac{1}{2} \qquad (6.10)$$

$$\int x^m e^{ax} dx = \frac{1}{a}x^m e^{ax} - \frac{m}{a}\int x^{m-1} e^{ax} dx \qquad (6.11)$$

$$\int \sin^m x \; dx = -\frac{\sin^{m-1} x \cos x}{m} + \frac{m-1}{m} \int \sin^{m-2} x \; dx \qquad (6.12)$$

$$\int \cos^m x \; dx = \frac{\cos^{m-1} x \sin x}{m} + \frac{m-1}{m} \int \cos^{m-2} x \; dx \qquad (6.13)$$

$$\int \sin^m x \cos^n x \; dx = \frac{\sin^{m+1} x \cos^{n-1} x}{m+n} + \frac{n-1}{m+n} \int \sin^m x \cos^{n-2} x \; dx$$

$$= -\frac{\sin^{m-1} x \cos^{n+1} x}{m+n} + \frac{m-1}{m+n} \int \sin^{m-2} x \cos^n x \; dx, \; m \neq -n \qquad (6.14)$$

$$\int x^m \sin bx \; dx = -\frac{x^m}{b} \cos bx + \frac{m}{b} \int x^{m-1} \cos bx \; dx \qquad (6.15)$$

$$\int x^m \cos bx \; dx = \frac{x^m}{b} \sin bx - \frac{m}{b} \int x^{m-1} \sin bx \; dx \qquad (6.16)$$

Example 6.6 Find (a) $\int \dfrac{dx}{\left(1+x^2\right)^{5/2}}$ and (b) $\int \left(9+x^2\right)^{3/2} dx$.

(a) Since the exponent in the denominator can be reduced by 1, we use this formula twice to obtain

$$\int \frac{dx}{\left(1+x^2\right)^{5/2}} = \frac{x}{3\left(1+x^2\right)^{3/2}} + \frac{2}{3} \int \frac{dx}{\left(1+x^2\right)^{3/2}}$$

$$= \frac{x}{3\left(1+x^2\right)^{3/2}} + \frac{2}{3} \frac{x}{\left(1+x^2\right)^{1/2}} + C$$

(b) Using the appropriate reduction formula, we obtain

$$\int \left(9+x^2\right)^{3/2} dx = \frac{1}{4} x \left(9+x^2\right)^{3/2} + \frac{27}{4} \int \left(9+x^2\right)^{1/2} dx$$

$$= \frac{1}{4} x \left(9+x^2\right)^{3/2} + \frac{27}{4} \left[x \left(9+x^2\right)^{1/2} + 9 \ln \left(x + \sqrt{9+x^2}\right) \right] + C$$

Trigonometric Integrals

The following identities, including those found in Table 4-1, are employed to find some of the trigonometric integrals in this section:

1. $\sin^2 x + \cos^2 x = 1$

2. $1 + \tan^2 x = \sec^2 x$

3. $1 + \cot^2 x = \csc^2 x$

4. $\sin^2 x = \dfrac{1}{2}(1 - \cos 2x)$

5. $\cos^2 x = \dfrac{1}{2}(1 + \cos 2x)$

6. $\sin x \cos x = \dfrac{1}{2}\sin 2x$

7. $\sin x \cos y = \dfrac{1}{2}[\sin(x-y) + \sin(x+y)]$

8. $\sin x \sin y = \dfrac{1}{2}[\cos(x-y) - \cos(x+y)]$

9. $\cos x \cos y = \dfrac{1}{2}[\cos(x-y) + \cos(x+y)]$

10. $1 - \cos x = 2\sin^2\dfrac{1}{2}x$

11. $1 + \cos x = 2\cos^2\dfrac{1}{2}x$

12. $1 \pm \sin x = 1 \pm \cos\left(\dfrac{1}{2}\pi - x\right)$

Two special substitution rules are useful in a few simple cases:

1. For $\int \sin^m x \cos^n x\, dx$: If m is odd, substitute u = cos x. If n is odd, substitute u = sin x.

2. For $\int \tan^m x \sec^n x \, dx$: If n is even, substitute u = tan x. If m is odd, substitute u = sec x.

Example 6.7 Evaluate the integral $\int \sin^2 x \, dx$.

$$\int \sin^2 x \, dx = \int \frac{1}{2}(1 - \cos 2x) \, dx = \frac{1}{2}x - \frac{1}{4}\sin 2x + C$$

Trigonometric Substitutions

Some integrations may be simplified with the following substitutions:

1. If an integrand contains $\sqrt{a^2 - x^2}$, substitute x = a sin z.

2. If an integrand contains $\sqrt{a^2 + x^2}$, substitute x = a tan z.

3. If an integrand contains $\sqrt{x^2 - a^2}$, substitute x = a sec z.

More generally, an integrand that contains one of the forms

$$\sqrt{a^2 - b^2x^2}, \quad \sqrt{a^2 + b^2x^2}, \quad \text{or} \quad \sqrt{b^2x^2 - a^2}$$

but no other irrational factor may be transformed into another involving trigonometric functions of a new variable as follows:

For	Use	To obtain
$\sqrt{a^2 - b^2x^2}$	$x = \dfrac{a}{b}\sin z$	$a\sqrt{1 - \sin^2 z} = a\cos z$
$\sqrt{a^2 + b^2x^2}$	$x = \dfrac{a}{b}\tan z$	$a\sqrt{1 + \tan^2 z} = a\sec z$
$\sqrt{b^2x^2 - a^2}$	$x = \dfrac{a}{b}\sec z$	$a\sqrt{\sec^2 z - 1} = a\tan z$

Table 6-1

In each case, integration yields an expression in the variable z. The corresponding expression in the original variable may be obtained by the use of a right triangle as shown in the following example.

Example 6.8 Find $\int \dfrac{dx}{x^2 \sqrt{4+x^2}}$

Let $x = 2 \tan z$, so that x and z are related as in Figure 6-1. Then

$dx = 2 \sec^2 z\, dz$ and $\sqrt{4+x^2} = 2 \sec z$, and

$$\int \frac{dx}{x^2 \sqrt{4+x^2}} = \int \frac{2 \sec^2 z\, dz}{\left(4 \tan^2 z\right)\left(2 \sec z\right)} = \frac{1}{4} \int \frac{\sec z}{\tan^2 z}\, dz$$

$$= \frac{1}{4} \int \sin^{-2} z \cos z\, dz = -\frac{1}{4 \sin z} + C = -\frac{\sqrt{4+x^2}}{4x} + C$$

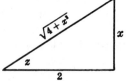

Figure 6-1

Integration by Partial Fractions

A polynomial in x as a function of the form

$$a_0 x^n + a_1 x^{n-1} + \bullet \bullet \bullet + a_{n-1} x + a_n,$$

where the a's are constants, $a_0 \neq 0$, and n, called the **degree** of the polynomial, is a nonnegative integer.

Every polynomial with real coefficients can be expressed (at least, theoretically) as a product of real linear factors of the form $ax + b$ and real irreducible quadratic factors of the form $ax^2 + bx + c$. (A polynomial of degree 1 or greater is said to be **irreducible** if it cannot be

> # Remember!
>
> If two polynomials of the same degree are equal for all values of the variable, then the coefficients of the like powers of the variable in the two polynomials are equal.

factored into polynomials of lower degree.) By the quadratic formula, $ax^2 + bx + c$ is irreducible if and only if $b^2 - 4ac < 0$. (In that case, the roots of $ax^2 + bx + c = 0$ are not real.)

Example 6.9

(a) $x^2 - x + 1$ is irreducible, since $(-1)^2 - 4(1)(1) = -3 < 0$.

(b) $x^2 - x - 1$ is not irreducible, since $(-1)^2 - 4(1)(-1) = 5 > 0$
   In fact,

$$x^2 - x - 1 = \left(x - \frac{1+\sqrt{5}}{2}\right)\left(x - \frac{1-\sqrt{5}}{2}\right)$$

A function $F(x) = f(x)/g(x)$, where $f(x)$ and $g(x)$ are polynomials, is called a **rational fraction**. If the degree of $f(x)$ is less than the degree of $g(x)$, $F(x)$ is called **proper**; otherwise, $F(x)$ is called **improper**.

An improper rational fraction can be expressed as the sum of a polynomial and a proper rational fraction. Thus,

$$\frac{x^3}{x^2+1} = x - \frac{x}{x^2+1}$$

Every proper rational fraction can be expressed (at least, theoretically) as a sum of simpler fractions (**partial fractions**) whose denominators are of the form $(ax + b)^n$ and $(ax^2 + bx + c)^n$, n being a positive integer.

Four cases, depending upon the nature of the factors of the denominator, arise.

Case I: Distinct Linear Factors

To each linear factor ax + b occurring once in the denominator of a proper rational fraction, there corresponds a single partial fraction of the form

$$\frac{A}{ax+b}$$

where A is a constant to be determined by the solution of simultaneous equations.

Example 6.10 Find $\int \frac{dx}{x^2-4}$.

We factor the denominator into (x - 2)(x + 2) and write

$$\frac{1}{x^2-4}=\frac{A}{x-2}+\frac{B}{x+2}$$

Clearing of fractions yields:

$$1 = A(x + 2) + B(x - 2) \tag{6-17}$$

or $\quad 1 = (A + B)x + (2A - 2B) \tag{6-18}$

We can determine the constants by either of two methods.

General Method: Equate coefficients of like powers of x in Eq. (6-18) and solve simultaneously for the constants. Thus,

A + B = 0 and 2A - 2B = 1

This yields

$$A = \frac{1}{4} \text{ and } B = -\frac{1}{4}$$

Short Method: Substitute in Eq. (6-17) the values x = 2 and x = - 2 to obtain 1 = 4A and 1 = - 4B; then

$$A = \frac{1}{4} \text{ and } B = -\frac{1}{4}$$

By either method, we have

$$\frac{1}{x^2-4} = \frac{\frac{1}{4}}{x-2} - \frac{\frac{1}{4}}{x+2}$$

Then,

$$\int \frac{dx}{x^2-4} = \frac{1}{4} \int \frac{dx}{x-2} - \frac{1}{4} \int \frac{dx}{x+2}$$

$$= \frac{1}{4} \ln|x-2| - \frac{1}{4} \ln|x+2| + C$$

$$= \frac{1}{4} \ln\left|\frac{x-2}{x+2}\right| + C$$

Case II: Repeated Linear Factors

To each linear factor $ax + b$ occurring n times in the denominator of a proper rational fraction, there corresponds a sum of n partial fractions of the form:

$$\frac{A_1}{ax+b} + \frac{A_2}{(ax+b)^2} + \cdots + \frac{A_n}{(ax+b)^n}$$

where the A's are constants to be determined.

Example 6.11 Find $\int \frac{(3x+5)dx}{x^3-x^2-x+1}$

$x^3 - x^2 - x + 1 = (x + 1)(x - 1)^2$. Hence,

$$\frac{3x+5}{x^3-x^2-x+1} = \frac{A}{x+1} + \frac{B}{x-1} + \frac{C}{(x-1)^2}$$

and

$$3x + 5 = A(x - 1)^2 + B(x + 1)(x - 1) + C(x + 1)$$

For $x = -1$, $2 = 4A$, and $A = 1/2$. For $x = 1$, $8 = 2C$ and $C = 4$. To determine the remaining constant, we use any other value of x, say $x = 0$; for $x = 0$, $5 = A - B + C$ and $B = -1/2$. Thus,

$$\int \frac{3x+5}{x^3-x^2-x+1} \, dx = \frac{1}{2} \int \frac{dx}{x+1} - \frac{1}{2} \int \frac{dx}{x-1} + 4 \int \frac{dx}{(x-1)^2}$$

$$= \frac{1}{2} \ln|x+1| - \frac{1}{2} \ln|x-1| - \frac{4}{x-1} + C$$

$$= -\frac{4}{x-1} + \frac{1}{2} \ln\left|\frac{x+1}{x-1}\right| + C$$

Case III: Distinct Quadratic Factors

To each irreducible quadratic factor $ax^2 + bx + c$ occurring once in the denominator of a proper rational fraction, there corresponds a single partial fraction of the form

$$\frac{Ax+B}{ax^2+bx+c}$$

where A and B are constants to be determined.

Example 6.12 Find $\int \frac{x^3+x^2+x+2}{x^4+3x^2+2} \, dx$.

$x^4 + 3x^2 + 2 = (x^2 + 1)(x^2 + 2)$. We write

$$\frac{x^3+x^2+x+2}{x^4+3x^2+2} = \frac{Ax+B}{x^2+1} + \frac{Cx+D}{x^2+2}$$

and obtain

$$x^3 + x^2 + x + 2 = (Ax + B)(x^2 + 2) + (Cx + D)(x^2 + 1)$$
$$= (A + C)x^3 + (B + D)x^2 + (2A + C)x + (2B + D)$$

Hence $A + C = 1$, $B + D = 1$, $2A + C = 1$, and $2B + D = 2$. Solving simultaneously yields $A = 0$, $B = 1$, $C = 1$, $D = 0$. Thus,

$$\int \frac{x^3+x^2+x+2}{x^4+3x^2+2}\,dx = \int \frac{dx}{x^2+1} + \int \frac{x\,dx}{x^2+2} = \arctan x + \frac{1}{2}\ln\left(x^2+2\right) + C$$

Case IV: Repeated Quadratic Factors

To each irreducible quadratic factor $ax^2 + bx + c$ occurring n times in the denominator of a proper rational fraction, there corresponds a sum of n partial fractions of the form:

$$\frac{A_1 x + B_1}{ax^2+bx+c} + \frac{A_2 x + B_2}{\left(ax^2+bx+c\right)^2} + \cdots + \frac{A_n x + B_n}{\left(ax^2+bx+c\right)^n}$$

where the A's and B's are constants to be determined.

Example 6.13 Find

$$\int \frac{x^5-x^4+4x^3-4x^2+8x-4}{\left(x^2+2\right)^3}\,dx$$

We write

$$\frac{x^5-x^4+4x^3-4x^2+8x-4}{\left(x^2+2\right)^3} = \frac{Ax+B}{x^2+2} + \frac{Cx+D}{\left(x^2+2\right)^2} + \frac{Ex+F}{\left(x^2+2\right)^3}$$

Then

$$\begin{aligned}
x^5-x^4+4x^3-4x^2+8x-4 &= \left(Ax+B\right)\left(x^2+2\right)^2 + \left(Cx+D\right)\left(x^2+2\right) + Ex+F \\
&= Ax^5 + Bx^4 + \left(4A+C\right)x^3 + \left(4B+D\right)x^2 \\
&\quad + \left(4A+2C+E\right)x + \left(4B+2D+F\right)
\end{aligned}$$

from which $A = 1$, $B = -1$, $C = 0$, $D = 0$, $E = 4$, $F = 0$. Thus, the given integral is equal to:

$$\int \frac{x-1}{x^2+2}\,dx + 4\int \frac{x}{\left(x^2+2\right)^3}\,dx =$$

$$\frac{1}{2}\ln\,(x^2+2) - \frac{\sqrt{2}}{2}\arctan\frac{x}{\sqrt{2}} - \frac{1}{\left(x^2+2\right)^2} + C$$

Miscellaneous Substitutions

If an integrand is rational except for a radical of the form

1. $\sqrt[n]{ax+b}$, then the substitution $ax + b = z^n$ will replace it with a rational integrand.

2. $\sqrt{q+px+x^2}$, then the substitution $q + px + x^2 = (z - x)^2$ will replace it with a rational integrand.

3. $\sqrt{q+px-x^2} = \sqrt{(\alpha+x)(\beta-x)}$, then the subsitution $q + px - x^2 = (\alpha + x)^2 z^2$ or $q + px - x^2 = (\beta - x)^2 z^2$ will replace it with a rational integrand.

Example 6.14 Find

$$\int \frac{dx}{x\sqrt{1-x}}\,.$$

Let $1 - x = z^2$. Then $x = 1 - z^2$, $dx = -2z\,dz$, and

$$\int \frac{dx}{x\sqrt{1-x}} = \int \frac{-2z\,dz}{(1-z^2)z} = -2\int \frac{dz}{1-z^2}$$

$$= -\ln\left|\frac{1+z}{1-z}\right| + C = \ln\left|\frac{1-\sqrt{1-x}}{1+\sqrt{1-x}}\right| + C$$

Other Substitutions

The substitution $x = 2 \arctan z$ will replace any rational function of $\sin x$ and $\cos x$ with a rational function of z, since

$$\sin x = \frac{2z}{1+z^2} \; ; \; \cos x = \frac{1-z^2}{1+z^2} \; ; \; \text{and } dx = \frac{2\,dz}{1+z^2}$$

The first and second of these relations are obtained from Figure 6-2, and the third by differentiating $x = 2 \arctan z$. After integrating, use

$z = \tan\dfrac{1}{2}x$ to return to the original variable.

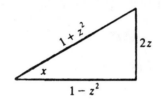

Figure 6-2

Example 6.15 Evaluate the following integral

$$\int \frac{dx}{1+\sin x - \cos x} = \int \frac{\dfrac{2\,dz}{1+z^2}}{1 + \dfrac{2z}{1+z^2} - \dfrac{1-z^2}{1+z^2}} = \int \frac{dz}{z(1+z)}$$

$$= \ln|z| - \ln|1+z| + C = \ln\left|\frac{z}{1+z}\right| + C = \ln\left|\frac{\tan\dfrac{1}{2}x}{1+\tan\dfrac{1}{2}x}\right| + C$$

Integration of Hyperbolic Functions

The following formulas are direct consequences of the differentiation formulas of Chapter 3.

$$\int \sinh x \, dx = \cosh x + C$$

$$\int \cosh x \, dx = \sinh x + C$$

$$\int \tanh x \, dx = \ln \cosh x + C$$

$$\int \coth x \, dx = \ln |\sinh x| + C$$

$$\int \operatorname{sech}^2 x \, dx = \tanh x + C$$

$$\int \operatorname{csch}^2 x \, dx = -\coth x + C$$

$$\int \operatorname{sech} x \tanh x \, dx = -\operatorname{sech} x + C$$

$$\int \operatorname{csch} x \coth x \, dx = -\operatorname{csch} x + C$$

$$\int \frac{dx}{\sqrt{x^2 + a^2}} = \sinh^{-1} \frac{x}{a} + C$$

$$\int \frac{dx}{\sqrt{x^2 - a^2}} = \cosh^{-1} \frac{x}{a} + C, \ x > a > 0$$

$$\int \frac{dx}{a^2 - x^2} = \frac{1}{a} \tanh^{-1} \frac{x}{a} + C, \ x^2 < a^2$$

$$\int \frac{dx}{x^2 - a^2} = -\frac{1}{a} \coth^{-1} \frac{x}{a} + C, \ x^2 > a^2$$

Example 6.16 Evaluate the following integrals

$$\int \sinh \frac{1}{2}x \; dx \text{ and } \int \operatorname{sech}^2 (2x-1)dx$$

$$\int \sinh \frac{1}{2}x \; dx = 2 \int \sinh \frac{1}{2}x \; d\left(\frac{1}{2}x\right) = 2 \cosh \frac{1}{2}x + C$$

$$\int \operatorname{sech}^2 (2x-1)dx = \frac{1}{2}\int \operatorname{sech}^2 (2x-1)d(2x-1) = \frac{1}{2}\tanh(2x-1) + C$$

Applications of Indefinite Integrals

When the equation $y = f(x)$ of a curve is known, the slope m at any point $P(x, y)$ on it is given by $m = f'(x)$. Conversely, when the slope of a curve at a point $P(x, y)$ on it is given by $m = dy/dx = f'(x)$, a family of curves, $y = f(x) + C$, may be found by integration.

Remember!

To single out a particular curve of the family, it is necessary to assign or to determine a particular value of C. This may be done by prescribing that the curve pass through a given point. This is known as an initial condition.

Example 6.17 Find the equation of the family of curves whose slope at any point $P(x, y)$ is $m = 3x^2y$. Find the equation of the curve of the family which passes through the point (0, 8).

Since $m = \dfrac{dy}{dx} = 3x^2y$, we have $\dfrac{dy}{y} = 3x^2dx$. Then ln y = x^3 + C = x^3 +

ln c and $y = ce^{x^3}$. When $x = 0$ and $y = 8$, then $8 = ce^0 = c$. The equation of the required curve is $y = 8e^{x^3}$.

We can also use indefinite integrals to describe equations of motion. An equation $s = f(t)$, where s is the distance at time t of a body from a fixed point in its (straight-line) path, completely defines the motion of the body. The velocity and acceleration at time t are given by:

$$v = \frac{ds}{dt} = f'(t) \quad \text{and} \quad a = \frac{dv}{dt} = \frac{d^2s}{dt^2} = f''(t)$$

Conversely, if the velocity (or acceleration) is known at time t, together with the position (or velocity) at some given instant, usually at $t = 0$, the equation of motion may be obtained.

Example 6.18 A ball is rolled over a level lawn with initial velocity 25 feet per second (ft/sec). Due to friction, the velocity decreases at the rate of 6 ft/sec². How far with the ball roll?

Here $dv/dt = -6$. So $v = -6t + C_1$. When $t = 0$, $v = 25$; hence $C_1 = 25$ and $v = -6t + 25$.

Since $v = ds/dt = -6t + 25$, integration yields $s = -3t^2 + 25t + C_2$. When $t = 0$, $s = 0$; hence $C_2 = 0$ and $s = -3t^2 + 25t$.
When $v = 0$, $t = \frac{25}{6}$; hence the ball rolls for $\frac{25}{6}$ sec before coming to rest. In that time, it rolls a distance

$$s = -3\left(\frac{25}{6}\right)^2 + 25\left(\frac{25}{6}\right) = -\frac{625}{12} + \frac{625}{6} = \frac{625}{12} \text{ ft}$$

Solved Problems

Solved Problem 6.1 Evaluate the indefinite integral:

$$\int \frac{dx}{\sqrt[3]{x^2}}$$

Solution.

$$\int \frac{dx}{\sqrt[3]{x^2}} = \int x^{-\frac{2}{3}} dx = \frac{x^{\frac{1}{3}}}{\frac{1}{3}} + C = 3x^{\frac{1}{3}} + C$$

Solved Problem 6.2 Evaluate the indefinite integrals:

(a) $\int \tan x \, dx$ (b) $\int \tan 2x \, dx$

Solution.

(a) $\int \tan x \, dx = \int \frac{\sin x}{\cos x} dx = - \int \frac{-\sin x}{\cos x} dx$

$$= -\ln|\cos x| + C = \ln|\sec x| + C$$

(b) $\int \tan 2x \, dx = \frac{1}{2} \int (\tan 2x)(2 \, dx) = \frac{1}{2} \ln|\sec 2x| + C$

Solved Problem 6.3 Evaluate the integral

$$\int \sin^2 \theta \cos \theta \, d\theta$$

Solution. We begin by making the following definitions:

$u = \sin \theta$ \qquad or \qquad $du = \cos \theta \, d\theta$

Therefore,

$$\int \sin^2 \theta \cos \theta \, d\theta = \int u^2 \, du$$

$$= \frac{u^3}{3} + C$$

$$= \frac{\sin^3 \theta}{3} + C$$

Solved Problem 6.4 Evaluate the integral

$$\int \left(1 + x^3\right)^5 x^2 \, dx$$

Solution. We first begin by defining

$$u = 1 + x^3$$

It follows that

$$du = 3x^2 \, dx \quad \text{or} \quad \frac{du}{3} = x^2 \, dx$$

Thus,

$$\int \left(1 + x^3\right)^5 x^2 \, dx = \int u^5 \, \frac{du}{3}$$

$$= \frac{1}{3} \int u^5 du$$

$$= \frac{1}{3} \frac{u^6}{6} + C$$

$$= \frac{\left(1 + x^3\right)^6}{18} + C$$

Solved Problem 6.5 Evaluate the integral

$\int x \ln x\ dx$

Solution. We begin by defining

$u = \ln x$

which implies

$$d(\ln x) = \frac{dx}{x}$$

Thus,

$$\int x \ln x\ dx = \int (\ln x)(x\ dx)$$

where $u = \ln x$ and $dv = x\ dx$. Integrating by parts yields

$$\int x \ln x\ dx = \int (\ln x)(x\ dx)$$

$$= (\ln x)\left(\frac{x^2}{2}\right) - \int \left(\frac{x^2}{2}\right)\frac{dx}{x}$$

$$= \frac{x^2 \ln x}{2} - \int \frac{x\ dx}{2}$$

$$= \frac{x^2 \ln x}{2} - \frac{x^2}{4} + C$$

Solved Problem 6.6 Evaluate the integral

$\int x^2 e^x\ dx$

Solution. This is a problem requiring integration by parts. We begin by defining:

$u = x^2 \quad du = 2x\ dx$

$dv = e^x\ dx \qquad v = e^x$

$\int x^2 e^x\ dx = x^2 e^x - \int e^x\ 2x\ dx = x^2 e^x - 2\int x e^x\ dx$

The integral $\int x e^x\ dx$, which is also solved using integration by parts, is given by

$\int x e^x\ dx = x e^x - e^x$

Therefore,

$\int x^2 e^x\ dx = x^2 e^x - 2\left[x e^x - e^x \right] + C$

$$= x^2 e^x - 2x e^x + 2e^x + C$$

Solved Problem 6.7 Evaluate the integral

$\int_1^5 \dfrac{x^2 + 1}{2x - 3}\,dx$

Solution. We start with the substitution

$u = 2x - 3$

from which

$du = 2\ dx; \qquad dx = \dfrac{du}{2}; \qquad \text{and, } x = \dfrac{u+3}{2}$

Therefore,

$$\int \frac{x^2+1}{2x-3}\,dx = \int \frac{\left[\dfrac{(u+3)}{2}\right]^2 + 1}{u}\cdot \frac{du}{2}$$

$$= \int \frac{u^2+6u+13}{8u}\,du$$

$$= \int \left(\frac{u}{8}+\frac{3}{4}+\frac{13}{8u}\right)du$$

$$= \frac{u^2}{16}+\frac{3u}{4}+\frac{13}{8}\ln|u|+C$$

$$= \frac{(2x-3)^2}{16}+\frac{3}{4}(2x-3)+\frac{13}{8}\ln|2x-3|+C$$

Now that we have the general form for the solution of the integral, we can then evaluate over the defined limits to obtain the numerical value.

$$\int_1^5 \frac{x^2+1}{2x-3}\,dx = \left[\frac{(2x-3)^2}{16}+\frac{3}{4}(2x-3)+\frac{13}{8}\ln|2x-3|+C\right]_1^5$$

$$= \left[\frac{7^2}{16}+\frac{21}{4}+\frac{13}{8}\ln 7+C\right]-\left[\frac{(-1)^2}{16}-\frac{3}{4}+\frac{13}{8}\ln 1+C\right]$$

$$= 9+\frac{13}{8}\ln 7$$

Chapter 7
THE DEFINITE INTEGRAL, PLANE AREAS BY INTEGRATION, IMPROPER INTEGRALS

IN THIS CHAPTER:

✔ *Riemann Sums*
✔ *The Definite Integral*
✔ *Plane Areas by Integration*
✔ *Improper Integrals*
✔ *Solved Problems*

Riemann Sums

Let $a \leq x \leq b$ be an interval on which a given function $f(x)$ is continuous. Divide the interval into n subintervals $h_1, h_2, \ldots, h_n$ by the insertion of n - 1 points $\xi_1, \xi_2, \ldots, \xi_{n-1}$, where $a < \xi_1 < \xi_2 < \ldots < \xi_{n-1} < b$, and relabel a as ξ_0 and b as ξ_n. Denote the length of the subinterval h_1

by $\Delta_1 x = \xi_1 - \xi_0$, of h_2 by $\Delta_2 x = \xi_2 - \xi_1, \ldots$, of h_n by $\Delta_n x = \xi_n - \xi_{n-1}$. (This is done in Figure 7-1. The lengths are directed distances, each being positive in view of the above inequality.) On each subinterval, select a point (x_1 on the subinterval h_1, x_2 on $h_2, \ldots$, x_n on h_n) and form the Riemann sum

$$S_n = \sum_{k=1}^{n} f(x_k) \Delta_k x = f(x_1) \Delta_1 x + f(x_2) \Delta_2 x + \cdots + f(x_n) \Delta_n x \qquad (7.1)$$

each term being the product of the length of a subinterval and the value of the function at the selected point on that subinterval. Denote by λ_n the length of the longest subinterval appearing in Eq. (7.1). Now let the number of subintervals increase indefinitely in such a manner that $\lambda_n \to 0$. (One way of doing this would be to bisect each of the original subintervals, then bisect each of these, and so on.) Then

$$\lim_{n \to +\infty} S_n = \lim_{n \to +\infty} \sum_{k=1}^{n} f(x_n) \Delta_n x$$

exists and is the same for all methods of subdividing the interval $a \le x \le b$, so long as the condition $\lambda_n \to 0$ is met, and for all choices of the points x_k in the resulting subintervals.

Figure 7-1

The Definite Integral

Theorem: The **fundamental theorem of calculus** states the equivalence of the anti-derivative and the area under the curve over an interval [a, b] as:

$$\int_a^b f(x) \, dx = \lim_{n \to +\infty} S_n = \lim_{n \to +\infty} \sum_{k=1}^{n} f(x_k) \Delta_k x$$

The symbol $\int_a^b f(x)dx$ is read "the **definite integral** of f(x), with respect to x, from x = a to x = b." The function f(x) is called the **integrand**; a and b are called, respectively, the **lower** and **upper limits** (bounds) **of integration**. This rule establishes the inverse relationship between differentiation and integration. That is, if F'(x) = f(x) over [a, b], then

$$\int_a^b f(x)dx = F(x)\Big|_a^b = F(b) - F(a)$$

Example 7.1

(a) Take f(x) = c, a constant, and F(x) = cx; then

$$\int_a^b c\ dx = cx\Big|_a^b = c(b-a)$$

(b) Take f(x) = x and F(x) = $\dfrac{1}{2}x^2$; then

$$\int_0^5 x\ dx = \frac{1}{2}x^2\Big|_0^5 = \frac{25}{2} - 0 = \frac{25}{2}$$

We have defined $\int_a^b f(x)dx$ when a < b. The other cases are taken care of by the following definitions:

$$\int_a^a f(x)dx = 0$$

If a < b, then $\int_b^a f(x)dx = -\int_a^b f(x)dx$

Properties of Definite Integrals

If f(x) and g(x) are continuous on the interval of integration a ≤ x ≤ b, then

Property 7.1:

$$\int_a^b cf(x)\,dx = c\int_a^b f(x)\,dx, \text{ for any constant } c$$

Property 7.2:

$$\int_a^b [f(x)\pm g(x)]\,dx = \int_a^b f(x)\,dx \pm \int_a^b g(x)\,dx$$

Property 7.3:

$$\int_a^b f(x)\,dx = \int_a^c f(x)\,dx + \int_c^b f(x)\,dx, \text{ for } a < c < b$$

Property 7.4 (the first mean-value theorem):

$$\int_a^b f(x)\,dx = (b-a)f(x_0)$$

for at least one value $x = x_0$ between a and b. This also can be interpreted as a method of computing the *average value* of a continuous function over an interval [a, b], usually written in the form:

$$\overline{f(x)} = \frac{1}{b-a}\int_a^b f(x)\,dx$$

Property 7.5:

If $F(u) = \int_a^u f(x)\,dx$, then $\dfrac{d}{du}F(u) = f(u)$

The Theorem of Bliss

If f(x) and g(x) are continuous on the interval $a \le x \le b$, if the interval is divided into subintervals as before, and if two points are selected in each subinterval (that is, x_k and X_k in the kth subinterval), then

$$\lim_{n \to +\infty} \sum_{k=1}^{n} f(x_k) g(X_k) \Delta_k x = \int_a^b f(x) g(x) dx$$

We note first that the theorem is true if the points x_k and X_k are identical. The force of the theorem is that when the points of each pair are distinct, the result is the same as if they were coincident.

In evaluating definite integrals directly from the definition, we sometimes make use of the following summation formulas:

$$\sum_{k=1}^{n} k = 1 + 2 + \cdots + n = \frac{n(n+1)}{2}$$

$$\sum_{k=1}^{n} k^2 = 1^2 + 2^2 + \cdots + n^2 = \frac{n(n+1)(2n+1)}{6}$$

$$\sum_{k=1}^{n} k^3 = 1^3 + 2^3 + \cdots + n^3 = \left[\frac{n(n+1)}{2} \right]^2$$

Plane Areas By Integration

Area as the Limit of a Sum

If f(x) is continuous and nonnegative on the interval $a \le x \le b$, the Fundamental Theorem of Calculus allows us to identify the infinite Riemann sum with the definite integral

$$\int_a^b f(x) dx = \lim_{n \to +\infty} \sum_{k=1}^{n} f(x_k) \Delta_k x$$

can be given a geometric interpretation. Let the interval $a \le x \le b$ be subdivided and points x_k be selected as in the preceding section. Through each of the endpoints $\xi_o = a, \xi_1, \xi_2, \bullet \bullet \bullet, \xi_n = b$ erect perpendiculars to the x axis, and laterally by the abscissas x = a and x = b. Approximate each strip as a rectangle whose base is the lower base of the strip and whose altitude is the ordinate erected at the point x_k of the subinterval. The area of the kth approximating rectangle, shown in Figure 7-2, is

$f(x_k)\ \Delta_k x$. Hence $\sum\limits_{k=1}^{n} f(x_k)\Delta_k x$ is simply the sum of the areas of the n approximating rectangles.

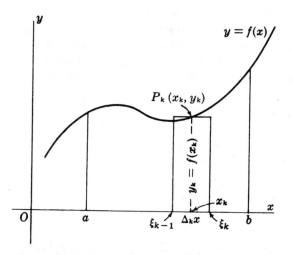

Figure 7-2

The limit of this sum is $\int\limits_{a}^{b} f(x)\ dx$; it is also, by definition, the area of the portion of the plane described above, or, more briefly, the area under the curve from x = a to x = b.

Similarly, if x = g(y) is continuous and nonnegative on the interval $c \leq y \leq d$, the definite integral $\int\limits_{c}^{d} g(y)\ dy$ is by definition the area bounded by the curve x = g(y), the y axis, and the ordinates y = c and y = d.

If $y = f(x)$ is continuous and nonpositive on the interval $a \leq x \leq b$, then $\int_a^b f(x)\, dx$ is negative, indicating that the area lies below the x axis.

Similarly, if $x = g(y)$ is continuous and nonpositive on the interval $c \leq y \leq d$, then $\int_c^d g(y)\, dy$ is negative, indicating that the area lies to the left of the y axis.

If $y = f(x)$ changes sign on the interval $a \leq x \leq b$, or if $x = g(y)$ changes sign on the interval $c \leq y \leq d$, then the area "under the curve" is given by the sum of two or more definite integrals.

Areas By Integration

The steps in setting up a definite integral that yields a required area are:

1. Make a sketch showing the area sought, a representative (kth) strip, and the approximating rectangle. We shall generally show the representative subinterval of length Δx (or Δy), with the point x_k (or y_k) on this subinterval as its midpoint.
2. Write the area of the approximating rectangle and the sum for the n rectangles.
3. Assume the number of rectangles to increase indefinitely, and apply the fundamental theorem of the preceding section.

Areas Between Curves

Assume that $f(x)$ and $g(x)$ are continuous functions such that $0 \leq g(x) \leq f(x)$ for $a \leq x \leq b$. Then the area A of the region R between the graphs of $y = f(x)$ and $y = g(x)$ and between $x = a$ and $x = b$ (see Figure 7-3) is given by

$$A = \int_a^b f(x)\,dx - \int_a^b g(x)\,dx = \int_a^b [f(x) - g(x)]\,dx$$

That is, the area A is the difference between the area $\int_a^b f(x)\,dx$ of the region above the x axis and below y = f(x) and the area $\int_a^b g(x)dx$ of the region above the x axis and below y = g(x).

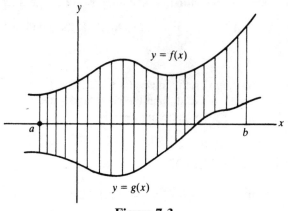

$$y = f(x)$$

$$y = g(x)$$

Figure 7-3

Example 7.2 Find the area enclosed by the curve $y^2 = x^2 - x^4$.

The curve is symmetric with respect to the coordinate axes. Hence the required area is four times the portion lying in the first quadrant.

For the approximating rectangle shown in Figure 7-4, the width is Δx, the height is $y = \sqrt{x^2 - x^4} = x\sqrt{1 - x^2}$, and the area is $x\sqrt{1 - x^2}\,\Delta x$ Hence the required area is

$$A = 4\int_0^1 x\sqrt{1 - x^2}\,dx = \left[-\frac{4}{3}\left(1 - x^2\right)^{3/2}\right]_0^1 = \frac{4}{3} \text{ square units}$$

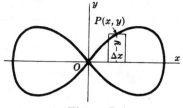

$$P(x, y)$$

Figure 7-4

Improper Integrals

The definite integral $\int_a^b f(x)\,dx$ is called an **improper integral** if either

1. The integrand f(x) has one or more points of discontinuity on the interval $a \le x \le b$, or
2. At least one of the limits of integration is infinite.

Discontinuous Integrand

If f(x) is continuous on the interval $a \le x < b$ but is discontinuous at $x = b$, we define

$$\int_a^b f(x)\,dx = \lim_{\varepsilon \to 0^+} \int_a^{b-\varepsilon} f(x)\,dx \text{ provided the limit exists}$$

(This can also be written $\lim_{\beta \to b} \int_a^\beta f(x)\,dx$.)

If f(x) is continuous on the interval $a < x \le b$ but is discontinuous at $x = a$, we define

$$\int_a^b f(x)\,dx = \lim_{\varepsilon \to 0^+} \int_{a+\varepsilon}^b f(x)\,dx \text{ provided the limit exists}$$

which can also be found as $\lim_{\alpha \to a} \int_\alpha^b f(x)\,dx$.

If f(x) is continuous for all values of x on the interval $a \le x \le b$ except at $x = c$, where $a < c < b$, we define

$$\int_a^b f(x)\,dx = \lim_{\varepsilon \to 0^+} \int_a^{c-\varepsilon} f(x)\,dx + \lim_{\varepsilon \to 0^+} \int_{c+\varepsilon}^b f(x)\,dx \text{ provided both limits exist.}$$

Example 7.3 Show that $\int_0^2 \dfrac{dx}{2-x}$ is meaningless.

The integrand is discontinuous at x = 2. We consider

$$\lim_{\varepsilon \to 0^+} \int_0^{2-\varepsilon} \frac{dx}{2-x} = \lim_{\varepsilon \to 0^+} \left[\ln \frac{1}{2-x} \right]_0^{2-\varepsilon} = \lim_{\varepsilon \to 0^+} \left(\ln \frac{1}{\varepsilon} - \ln \frac{1}{2} \right)$$

The limit does not exist; so the integral is meaningless.

Infinite Limits of Integration

If f(x) is continuous on every interval a ≤ x ≤ U, we define

$$\int_a^{+\infty} f(x)\,dx = \lim_{U \to +\infty} \int_a^U f(x)\,dx \quad \text{provided the limit exists}$$

If f(x) is continuous on every interval u ≤ x ≤ b, we define

$$\int_{-\infty}^b f(x)\,dx = \lim_{u \to -\infty} \int_u^b f(x)\,dx \quad \text{provided the limit exists.}$$

If f(x) is continuous, we define

$$\int_{-\infty}^{+\infty} f(x)\,dx = \lim_{U \to +\infty} \int_a^U f(x)\,dx + \lim_{u \to -\infty} \int_u^a f(x)\,dx$$

provided **both** limits exist.

Example 7.4 Evaluate $\int_0^{+\infty} \dfrac{dx}{x^2+4}$.

The upper limit of integration is infinite. We consider

$$\lim_{U \to +\infty} \int_0^U \frac{dx}{x^2+4} = \lim_{U \to +\infty} \left[\frac{1}{2} \arctan \frac{1}{2} x \right]_0^U = \frac{\pi}{4}$$

from which

$$\int_0^{+\infty} \frac{dx}{x^2+4} = \frac{\pi}{4}$$

Solved Problems

Solved Problem 7.1 Given the region bounded by the curve y = x^2, the line y = $-\dfrac{1}{2}x$, and the line x = 3, find the area of the region.

Solution. The area between two curves can be determined by:

$$area = \int_a^b \left[f(x) - g(x) \right] dx$$

where, in this case,

$f(x) = x^2$ and $g(x) = -\dfrac{1}{2}x$. Therefore, the area of this region is:

$$
\begin{aligned}
area &= \int_0^3 \left[x^2 - \left(-\frac{1}{2}x \right) \right] dx \\
&= \left(\frac{x^3}{3} + \frac{x^2}{4} \right) \Bigg|_0^3 \\
&= \left[\frac{3^3}{3} + \frac{3^2}{4} \right] - \left[\frac{0^3}{3} + \frac{0^2}{4} \right] \\
&= \frac{27}{3} + \frac{9}{4} \\
&= \frac{135}{12} \\
&= 11.25
\end{aligned}
$$

Appendix A
DIFFERENTIATION FORMULAS FOR COMMON MATHEMATICAL FUNCTIONS

Polynomial Functions

$$\frac{d}{dx}(a_o x^n) = n\, a_o\, x^{n-1}$$

$$\frac{d}{dx}(a_o u^n) = n\, a_o\, u^{n-1}\, [\frac{du}{dx}]$$

Trigonometric Functions

$$\frac{d}{dx}(\sin u) = \cos u\, [\frac{du}{dx}]$$

$$\frac{d}{dx}(\cos u) = -\sin u\, [\frac{du}{dx}]$$

$$\frac{d}{dx}(\tan u) = \sec^2 u\, [\frac{du}{dx}]$$

$$\frac{d}{dx}(\cot u) = -\csc^2 u \left[\frac{du}{dx}\right]$$

$$\frac{d}{dx}(\sec u) = -\sec u \tan u \left[\frac{du}{dx}\right]$$

$$\frac{d}{dx}(\csc u) = -\csc u \cot u \left[\frac{du}{dx}\right]$$

$$\frac{d}{dx}(a^u) = a^u \ln a$$

Exponential Functions

$$\frac{d}{dx}(e^u) = e^u \left[\frac{du}{dx}\right]$$

Logarithmic Functions

$$\frac{d}{dx}(\ln u) = \frac{1}{u}\frac{du}{dx}$$

$$\frac{d}{dx}(\log_a u) = \frac{\log_a e}{u}\left[\frac{du}{dx}\right], \quad a \neq 0, 1$$

Appendix B
INTEGRATION FORMULAS
FOR COMMON
MATHEMATICAL FUNCTIONS

Polynomial Functions

$$\int u^p du = \frac{u^{p+1}}{p+1}, \quad p \neq 1$$

$$\int u^{-1} du = \int \frac{du}{u} = \ln u$$

Trigonometric Functions

$$\int \sin u \, du = -\cos u$$

$$\int \cos u \, du = \sin u$$

$$\int \tan u \, du = -\ln \cos u$$

$$\int \cot u \, du = \ln \sin u$$

$$\int \sec u \, du = \ln(\sec u + \tan u)$$

$$\int \csc u \, du = \ln(\csc u - \cot u)$$

Exponential Functions

$$\int a^u du = \frac{a^u}{\ln a}, \quad a > 0, a \neq 1$$

$$\int e^u \, du = e^u$$

Logarithmic Functions

$$\int \ln x \, dx = x \ln x - x$$

Index